Planet Hydrogen – The Taming of the Proton

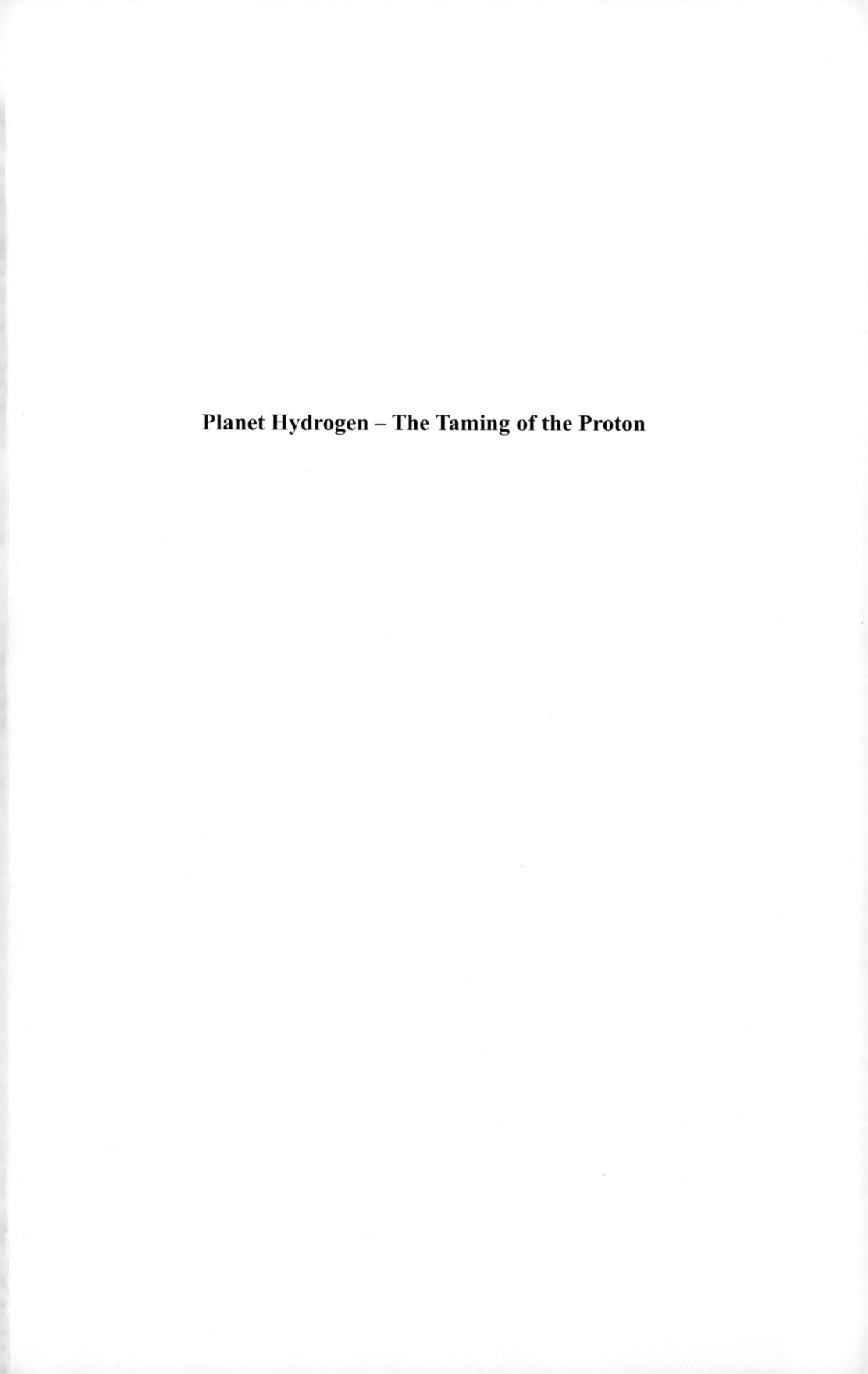

ACKNOWLEDGEMENTS

Bragi Árnason, Carl-Jochen Winter, Leó Kristjánsson, Baldur Arnarson, Chul-Hee Han, Harris Schaer, Runólfur Þórðarson, Ingibjörg Sigurðardóttir, Einar H. Guðmundsson, Maria Skyllas-Kazacos, Nicholas Vanderborgh, Stephenie Ritchie, Jón Björn Skúlason, María Maack, Hjalti Páll Ingólfsson, Hallmar Halldórs, Kristján Leósson, Hannes Jónsson, Volodymyr Yartys, Ingimundur Sigfússon, Stanislav Malyshenko, Boris Reutov, Bill David, Helgi Ágústsson, Bob Dixon, David Garman, Tryggve Folkeson, Joseph Maceda, Guðrún Pétursdóttir, Davíð Þór Þorsteinsson, Dagrún Inga Þorsteinsdóttir, Þorkell Viktor Þorsteinsson, David Hart, Ragnar Baldursson, Peter Edwards, Gunnar Ö. Gunnarsson, Steve Chalk, Úlfar Steindórsson, Hanns-Joachim Neef, Nejat Veziroglu, Frano Barbir, Line Hagen, Anthony Clemens, Mike Mills, Gunnar Pálsson, Víglundur Þór Víglundsson, Ingólfur Þorbjörnsson, Michael Cummings, Bjarni Bjarnason, Hallgrímur Jónasson, Marieke Rialt, Albert Albertsson, Þorleifur Finnsson, Helgi Þór Ingason, Þórir Ibsen, Rene Biasone, Victor Ivanovich Tatarintsev, Paulo Emilio Miranda, Wolfgang Dönitz, Juergen Garche, Hiro Tokushige, Kristín Ingólfsdóttir, The University of Iceland, The Innovation Center Iceland, Menningarsjóður: The Icelandic Cultural Foundation.

The author wishes to especially acknowledge:

Rósa Ólafsdóttir, B.Sc. for work on technical drawings.

Sigrún Sigvaldadóttir, Hunang Design, for the design of the cover.

Ragnar Axelsson, RAX, for the photograph on the backcover.

Front cover:
The idea behind the front cover picture is to show the Earth going through a transition. The Earth and the Moon symbolise the proton and the electron respectively. When the hydrogen atom approaches a proton exchange membrane, symbolised by the shady region on the picture, it leaves the electron on the conducting membrane surface and travels alone through the membrane in order to recombine with oxygen on the other side. This is the essence of the proton exchange membrane fuel cell.

For Bergthora

BIANCA
> *"The taming school?"*

TRANIO
> ***"Ay, mistress, and Petruchio is the master,***
> ***That teacheth tricks eleven and twenty long***
> ***To tame a shrew ..."***

William Shakespeare
The Taming of the Shrew
Act 4. Sc 2

CONTENTS

ACKNOWLEDGEMENTS v
PREFACE xiii

PART I – PLANET HYDROGEN
  A Letter from Casablanca 3

PART II – INTRODUCTION TO THE SCENE
  Showers from the Sun 9
  Low on petrol 14
  Exhaustion 17
  Burning and converting 22
    Thermodynamics: the flow from hot to cold 22
  Non-renewable types of available energy sources 25
    Coal 25
    Nuclear energy 27
  Renewable energy sources 28
    Photovoltaics 28
    Concentrating solar power 30
    Wind energy 31
    Hydroelectric power 31
    Biomass and Biofuels 33
    Geothermal energy 35
    Wave and tidal power 38
    Exotic future methods 39
  Decarbonisation and cleaner energy 40
  Sharpening the pencil 45

PART III – THE TAMING OF THE PROTON
  Discovering the proton 49
  Some key properties of hydrogen 52
  PRODUCING HYDROGEN 54
  Hydrogen from fossil-fuels and chemical processes 54
  Hydrogen from biomass 57
  Biological methods of producing hydrogen 58
  Hydrogen from nuclear energy 60
  "Jules Verne´s method" 61
  Hydrogen from wind 64
  Solar hydrogen 65
  Mimicking the Maple Leaf 67
  Hydrogen from geothermal vents 69
  STORING HYDROGEN 71
  Introduction 71
  Storage in the form of gas 71

Liquefaction and liquid storage 73
Solid state storage 75
Sorption properties 76
Binary systems and beyond 79
Complex hydrides 80
Alanates 80
Borohydrides 81
Exotic Hydrogen Storage Compounds 82
The Grand Challenge of the modern day alchemists 83
EFFICIENT HYDROGEN UTILISATION 84
Utilising the energy of the burning of hydrogen 84
The ultimate taming: The discovery and development of the fuel cell 86
The fuel cell menu 90
Proton Exchange Membrane Fuel Cells 93
A discussion of fuel cells and applications 94
Phosphoric Acid Fuel Cells 96
Alkaline Fuel Cells 97
Molten Carbonate Fuel Cells 98
Solid Oxide Fuel Cells 99
Direct Methanol Fuel Cell 101

PART IV – HYDROGEN INFRASTRUCTURE AND SOCIETY

Hydrogen entering society 107
Infrastructure 108
The Spirit of Davis 110
The HyWays of the European roadmap 114
The rising sun of Tokyo Bay 116
High stakes in a HySociety 117
Hydrogen with Foresight 119
Hydrogen Education 121
Sophisticated Economics of Hydrogen Energy 122
Rules of the game: Hydrogen safety, codes and standards 124
  Tampering with the elements water and fire 124
  The unexpected steam-iron process 125
  Hydrogen safety 127
  Acceptance in society through Codes and Standards 130
THE INTERNATIONAL HYDROGEN MOVEMENT 131
International Hydrogen Energy Association and the
hydrogen family 131
The International Energy Agency Hydrogen Implementation
Agreements. The hydrogen expert assembly 134
The International Partnership for the Hydrogen Economy IPHE:
Governments United for Hydrogen 135

PART V – AROUND THE WORLD IN EIGHTEEN LEAPS

Australia, New Zealand and Antarctica and the concern for the planet 141

Planet Hydrogen

Brazil setting the pace for a continent 144
Canada, the impressive hub of hydrogen technology 146
China could leapfrog and lead the way 148
The European Commission and the impressive policy work 150
France – The post-Carnot era 152
Germany, masters of endurance 154
Iceland, perhaps the ideal testing forum 157
India, the second wildcard 166
Italy and the renaissance of hydrogen 167
Japan: protonics added to electronics 169
Korea joining the forefront 171
Norway and the Nordic pioneers 172
Russia and the increased strong commitment to hydrogen 177
United Kingdom, where much of the science was developed 180
United States, the pacemaker 182

PART VI – NEAR A JOURNEY'S END 191

REFERENCES AND FURTHER READING 195

GLOSSARY 203

CYBER-APPENDIX .... www.tamingoftheproton.com

INDEX 209

PREFACE

AT THE ONSET OF A JOURNEY TO HYTOPIA

During the past decade I have had the privilege to give lectures and seminars to audiences in most continents of the world. The subject has been HYDROGEN as the carrier part of a renewable energy system.

In my work within the International Partnership for the Hydrogen Economy, IPHE, it has been possible to witness the problems facing the energy utilisation of the world as well as to acquire a close insight into the various pathways of hydrogen production, distribution, storage, utilisation and socio-economics all around the globe.

I wish to share with the reader the experience from my own country of Iceland where a substantial movement towards creating a hydrogen economy has been taking place since around 1998. The project enjoys support from all sectors of our society, government, academia and industry together with a substantial public attention and participation.

The present book is the harvest of a quiet eye as well as a personal participation over decades of work at the domestic and international level and fruitful discussion with hundreds of colleagues, students and stakeholders. It is composed in the spirit of the international enthusiasm for this most fundamental of fuels. The intention is to give the reader an insight into hydrogen and the expected hydrogen economy by mutually sharing some basic aspects of hydrogen and its physics, chemistry and engineering. While the author strives to tell an informative story, a conscious attempt is made to avoid much technical jargon although it can not be avoided in all cases.

The book is intended for the technically-minded layperson wishing to gain insight into the new future of energy. We will attempt to balance science, engineering, sociological as well as environmental aspects.

The somewhat Shakespearean title "Taming of the proton" originates in the beautiful physics of fuel cells which not only transform the classical thinking about energy but will eventually replace one energy era with a new one. The proton nucleus of hydrogen is as important a player in fuel cells as is the electricity in motors. The proton has proven much more difficult to harness and control than the electron. Its taming continues to be tricky.

The title also refers to a certain vision I want to share with the reader: The past century was characterised by electricity and electronics, where the technological society has been conquered by electrons, electricity and, later, information technology. On the other hand, the present century likewise may become the century of the proton and the simplest atom that can be created by one proton, namely hydrogen. Utilisation of this source will be realised through the use of fuel cells. A new and exciting era is emerging.

This is expected to be a tough challenge with many obstacles which we shall neither hide nor camouflage. A sometimes quixotic past will have to be replaced with realism and a good deal of patience.

The book is organised in the following manner: As the book unfolds, the reader will be given an insight into the future vision of the hydrogen economy by a reference to a post-card letter written from North Africa in 2048. Then we turn back to the present to an evening journey in Iceland and while glancing at the night sky illuminated by Northern Lights we take a look at the past, the origins of our universe and examine the evolution that has led to the material world we now live in and of which hydrogen is a major constituent. We will study the nature of energy and various energy sources and the problems the present energy portfolio of humans has created.

In subsequent pages we will look at the discovery of the proton and the essentials of the hydrogen energy technology where history of scientific advancement will be woven into the narrative. The reader will get an insight into production, storage, infrastructure and utilisation of hydrogen as reflected in various world wide experiments with components of a hydrogen society.

Together we will tour the world and look into the various ways different countries are experimenting with and introducing hydrogen into their energy systems during the first decade of the new millennium. To do this we shed light upon the international cooperation in the field of hydrogen with as much up to date information as possible.

It is my sincere hope that through this book the reader will gain much more insight into the nature of the element of hydrogen and the possible role of this lightest of elements in shaping our future energy society.

The age of the proton has dawned.

PART I
PLANET HYDROGEN

A Letter from Casablanca

Casablanca, June 4th 2048

Dear dad!

We wish you were here with us. The trip from Iceland went quite well and the prototype cryojet took only three hours to reach the new airport in the Sahara. We enjoyed very good visibility on the way with almost cloud free skies over the North Sea, a good view over the Alps in the east, a cloudy Iberia peninsula and continued clearer skies as we passed the strait of Gibraltar.

North of Norway we could clearly see the Heimdal II oil field where enormous quantities of CO_2 have been pumped down into the ocean bottom to sequester the greenhouse gas and at the same time increase the yield of the field which has come to the end of its useful life.

The old oil platforms in the North Sea which stopped production some decades ago, now flourish with a new task; they harbour wind generators making hydrogen which is piped to land at various places in the UK. Some of them are still extracting the remaining natural gas and reforming it on site to hydrogen. We could see the magnificent wind generators from the air: we were told that each platform produced a one gigawatt equivalent of hydrogen. It is like having the equivalent of a nuclear power plant there in the ocean – only that it is purely renewable based. There was not much to be seen on land and most likely the hydrogen is piped to a distribution system which is partly the old natural gas network and partly a new installation.

It was noteworthy as we passed London to see the large estuary of the river Thames which, after the dramatic melting of the polar ice and the Greenland icecap, has drowned a considerable amount of land, mostly marshland on the north and south banks. On the banks of the Thames where we remember the large coal power plants we could see no smoke coming out; most likely the emissions are captured directly and the carbon sequestered in situ.

In the plane there was a Danish family sitting next to us who told us that the old airport at Kastrup was now abandoned because of the hectic flooding that took place in the wake of hurricane Jensine in 2044. We mentioned that when you were a student in

Copenhagen some 70 years ago, hurricanes only occured in the Gulf of Mexico and the most Denmark could expect was a deep low pressure resulting in a windy day or two. But with the increased sea temperature in summer in the Channel, a combined effect of a hurricane and lower air pressure produced flooding conditions in Denmark – enough to threaten the airport construction and the bridge over to Sweden. Also, due to the weakening of the branch of the Gulf Stream going to Norway and Iceland, the Channel temperature seems to have shown a very different behaviour in the 2040s. The conveyor belt of ocean currents has been dramatically weakened over the past decade.

What amazed us when flying over the western part of the Alps was that we could not see any snow-cover at all on the highest mountain tops. We remember that the UN alarm target of 500 ppm CO_2 in the atmosphere was reached last summer. A series of mild winters in the Alps has not only decreased their glaciers to virtually nothing but of course also left the old and famous skiing areas almost completely free of snow. Some of the old skiing areas now have these gigantic carbon dioxide collectors and an associated sequestration programme.

Anyway, we passed the continent of Europe and were soon heading for a landing in the new Sahara airport. We remember you had told us about the Sahara project already in 2006 where the plan was to harness solar energy and wind energy on the Morocco side of the Sahara. We read something about a German-Sahara project on solar hydrogen stemming back from as early as the 1980s.

Now, this has materialised in a whole new world! From the air one can see the huge lake that was created when the ocean water of the Atlantic was led into the new Sahara Sea called Sebkha Tah. The geological depression was about 50 metres below the sea level before the project began; after digging deeper, the Moroccans led a pipeline from the Atlantic coast into the gigantic hole. As the seawater flows in, it goes through a hydroelectric turbine system and produces GigaWatts of electricity which is partly used to desalinate seawater for the tourist villages around the Sebkha Tah Sea and partly to produce hydrogen for the infrastructure of the new lakeshore. From our hotel we could see in the south reflections from the heliostats which are harnessing sun energy for electricity production. The electricity is fed under the strait of Gibraltar and over to Sicily to the European grid in huge DC cable bundles. Because of the constant evaporation from the Sebkha Tah due to endless sunshine, the seawater flows in and the hydroelectric station is constantly producing electricity. One of the tourists here pointed out to us, that some of the molten Greenland ice could effectively be stored in the Saharan sea – but only a small fraction. The developers here are taking care of deposited salt on the bottom of the lake with some sort of a collection system and redepositing it in the Atlantic Ocean. In fact, in the end of the Sea nearest to the inlet of the Atlantic water, all sorts of fish are being cultivated and seem to be forming a new aqueous biohabitat. We have only tasted the Atlantic tuna caught west of the coast.

The projects here in North Africa have made the region one of the most energy intensive places on Earth. In addition to the traditional renewable harnessing systems, we went on a guided tour into a small valley where solar energy is converted into hydrogen by biomimetic methods. From the distance the power plant area has a distinct white- greyish colour which stems from the ruthenium metal which characterises the artificial "chlorophyll" which is at the heart of the biomimetic systems. The hydrogen produced has its origins in water which is fed from the desalination plant.

When we asked the tour guide about the use of nuclear fusion he told us that although having been proven commercially feasible in France, North Africa had enough solar energy to provide a more economical solution than the French one. This seems to be the "regionality" of hydrogen use; wind, geothermal, nuclear fusion and bioenergy; with solar in some areas of Earth being the most advantageous and perhaps the cheapest way possible worldwide. The tour guide, an elderly academic type, smiled and said that "the civilization here started with solar energy ages ago – and is now back to the second solar civilization". Is it not exactly what you said that your German friend, Carl-Jochen Winter always had been saying all along?!

On the coast close to Tarfaya in the north, there is a giant harbour for tankers transporting liquid hydrogen from the Sahara project to customers world wide. Mostly to the U.S. and Northern Europe.

We are staying at a hotel not far from the biomimetic power plant and enjoying the good weather here close to the new Sea while writing this detailed letter to you. We started it in Casablanca but finished it here in Sebkha City. The whole region is like an oasis with palm trees, date palms and the like. The most common are. Mediterranean Fan Palms - multi-trunk palm that grows to about three metres. Also Blue Palms - slow-growing palms with arching, silvery-blue feather-like fronds; and Guadalupe Fan Palms with light green fan-shaped fronds.

The hotel uses electricity from the grid for all possible purposes. The hotel transport fleet runs on hydrogen from the hotel's own fueling station which is based on electrolysis of water.

The air conditioning system is based on metal hydride driven reactors which cool the air as a part of their absorption-desorption system. This cooling effect is so powerful that the chef of the kitchen told us that the hotel restaurant could store sensitive food, such as frozen tuna at very low temperatures down to minus 70º Centigrade, ideal for keeping this delicate food fresh in the African heat.

Most of the cars here are either ToyGeM or DCendai but there are also the IndiFords and the new Chinese Lao Tses. All these car companies participated in the Addis Ababa – Casablanca Rally. The harsh conditions for no refueling over 1000 km between stations was easily met by the new combination of liquid hydrogen and hydrides tanks. One of the lightest cars was able to do the whole stretch without ever refuelling. We rented a ToyGeM and have enjoyed taking short trips in the vicinity of the Sebkha Sea.

The boats on the lake are mostly fuel-cell powered but there are a few sailing boats which utilise the breeze that is always felt. We noticed that the big ferry ship which connects Casablanca and Lisbon is totally hydrogen powered but has a large solar assisted electricity system. Again here the storage is based on the combined liquid hydrogen and hydrides. Most of the luxurious yachts in the harbour seemed either solely hydrogen based or had wind/solar hybrid systems.

What we find so interesting is that the hydrogen energy economy has opened up for the people here a lot of new opportunities. A country which half a century ago was among the developing areas of the world now has a thriving economy developed on a whole new energy base. Harnessing renewables and producing hydrogen has made this area so prosperous. The United Nations have assisted them through this important transition and is setting an example for other developing countries.

Yesterday, we met an old Moroccan who said he had met you in Sevilla at a conference on hydrogen from renewables in 2005!

We just wish you could have been with us here! Somehow all the predicted elements of a hydrogen energy economy seem to have been assembled here. It feels like a privilege to stay in this hydrogen oasis. And this is of course only an example of what is happening all over the planet.

We are transferring a load of videoshots with this letter over the internet so that you can enjoy it. By the way, happy birthday! We will be home in a week.

All the best for now,

Your children, David, Dagrun and Thorkell

PART II
INTRODUCTION TO THE SCENE

Showers from the Sun

I am driving home from the countryside north of Reykjavik. It is just after dusk in early winter and the snow-covered mountains glow in the moonlight on the eastern sky. Crisp ice crystals on the road reflect the headlights of the car. There are no clouds and the night ahead will presumably be cold.

As I turn towards Reykjavik amidst the mountains of Borgarfjordur I notice a string of green turquoise colors dancing in the sky; dancing as if they were surrounding the Pole star or performing a Swan Lake ballet of heavenly proportions. These are the Northern Lights, Aurora Borealis, one of nature´s most wonderful celestial events.

The Northern lights (fig 1.) originate from the Sun. They stem from hectic eruptions on the solar surface; outbursts of very hot plasmatic materials. The Sun is primarily composed of the most important element in the universe: ***Hydrogen***. It is the simplest of the elements known in the periodic table. Usually it consists of only one proton surrounded by an electron which is about 2000 times lighter than the proton.

Figure 1. Car with headlights on traveling under the Northern Lights. Photo: F. Holm.

In the interior of the Sun, protons fuse into heavier nuclei of helium, named after Helios, the Greek word for Sun. This elemental "factory" is one of the wonders of our universe and is only the first phase of a larger scale cosmic production plant for most of the elements we know. We will come to the nature of this elements factory a little later in our narrative.

The bursting of the hot plasma at the Sun's surface throws out vast quantities of protons, electrons, helium nuclei also called alpha particles and other ions. Each alpha-particle has two protons, the positively charged building blocks of matter, joined to a couple of chargeless neutrons which play an important role in balancing the nuclear forces. They are sputtered out in all directions from the Sun, mainly in the so called Coronal Mass Ejections, and will create particle showers on the Moon and the Earth. The showers are fundamentally nasty to life. Being composed of highly charged clouds they are fortunately mostly captured by the shielding magnetic field of Earth. A charged particle entering the magnetic field experiences a sideways force in proportionality to its speed and charge. When the protons from the Sun reach the Inner Van Allen belt of Earth (fig.2) they start spiralling around the imaginary "magnetic field lines". Spiralling charged particles are accelerated to an extent where physical laws demand emissions of light: photons. Most importantly the particles hit both oxygen and nitrogen in the Earth's atmosphere. They excite electron states in the oxygen and cause it to emit photons in the green part of the color spectrum. When the nitrogen is hit and excited, it glows in the red-violet spectrum. In summary: protons and electrons from the Sun are blown away to Earth and when they collide and spiral in the magnetic field, they cause the spectacular colorful Northern Lights.

The night is getting darker, the Moon has dipped behind the mountains and the sky is partly illuminated by the Northern Lights. As I concentrate on viewing the Pole star I can also see the illumination of the Galaxy to which our solar system belongs.

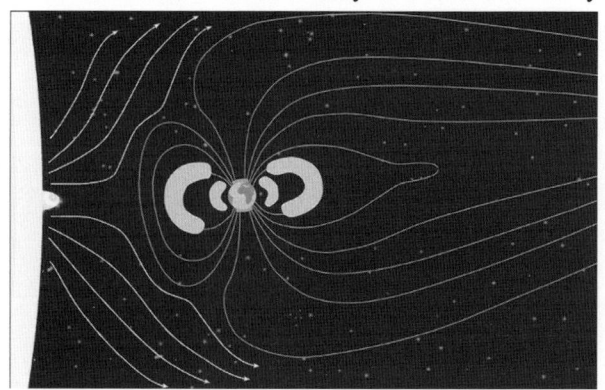

Figure 2. The Sun emits charged particles and the solar wind blows to the right of the picture. The Earth's magnetic field presents an obstacle to the solar wind, as would a rock in a running stream of water. This obstacle is called a "bow shock".
The bow shock slows down, heats, and compresses the solar wind, which then flows around the rest of Earth's magnetic field. The white doughnut shaped layers are the Van Allen Radiation Belts where a considerable amount of protons and other ions are stored. The regions above the poles are schematically showing the Auroras. (Based on NASA).

In my view to the sky right now I can assume that at least 92 per cent of the visible atoms is hydrogen. And this magnificent night-sky theatre also tells a very long story!

Cosmologists now assume that the universe was created about 13.7 billion years ago. The Poetic Eddas of Nordic mythology assume that the universe was created out of a void or "nothingness" called Ginnunga Gap. Modern science does in fact assume that this process, called the Big Bang, initiated in a super-dense and extremely hot state. The temperatures were so high that neither atoms nor nuclei could be formed. Perhaps it was only a mixture of so called "strings" intertwined in an unfathomable stirring soup of matter. A sub-atomic noodle soup!

The expansion of the universe had a cooling effect on its matter. As the soup cooled down a very short time after the initiation of the Big Bang, one can imagine that protons began to form from quarks and later pairs of protons formed and coupled to neutrons to form the aforementioned alpha particles.

As the temperature cooled off to 30 million degrees, one tenth of a second into the new time, it would be expected that the first protons would be formed as well as the neutrons and electrons. At only near 14 seconds after the Big Bang we would expect the first nuclei to emerge; the alpha-particles. Then finally, after a cooling for some 300 thousand years after the Big Bang, the first atoms would emerge from nuclei and electrons.

So, if a very short story of the hydrogen atom should be told, it would be close to the one we have just gone through. But what about the rest of the story of this remarkable creation? As I stare through my windshield on my way through the night I am well aware of the fact that streaming towards us is the light from distant stars, from remote parts of the cosmos, light photons that were emitted at very different stages in the history of the universe, even from the very early stage these 13.7 billion years ago.

Because of the nature of the binding forces of these elementary particles it turned out that protons in the early universe outnumbered alpha particles about eleven to one. So as the universe cooled it turned out that the mixture of the soup contained about 92 per cent hydrogen and 8 per cent helium (by number) and a fraction of a per cent of deuterium, a hydrogen atom with one neutron captured to share a place with a proton in its nucleus. So what was the role waiting for hydrogen in the rest of the process? It was indeed important.

The hydrogen atom winds itself like a red thread through the whole history of the evolution of the universe. The giant masses united in the stars, in the "suns", all begin the process of fusion into helium in their cores which are typically of Earth´s size. Then, depending upon the total mass, the fate of the original star will be decided. A common star evolution goes through the stage of the formation of heavier atoms. Hydrogen fuses into helium; Then carbon emerges, then oxygen, followed by silicon and finally iron in the most massive stars (our Sun, being relatively small, is expected to stop the nuclear fusion process around carbon).

Generally the fusion process enters crossroads in the case of the element iron and the fusion line stops. The star factory now comes to a new stage: Depending on total mass, gravity can become the dominant force and cause a collapse and crush the nuclear forces at the centre of the star. In such a scenario, which in fact only needs to exceed 1.4 times the mass of the stellar core of our Sun, the various elements can be thrown into space as a result of the stellar explosion to form the building blocks of planets with different forms and amounts of heavier elements. This fascinating saga goes a little beyond our focus and will not be followed in more detail. Let us only keep in mind that for a planet like Earth, a number of elements must have been present originally. Collisions with comets and meteorites can also have affected the composition of the original planet. As the Earth cooled down, hydrogen had the chance of uniting chemically with other elements to form compounds. The electron surrounding the hydrogen nucleus is very virile in making contacts with other elements. Hence, chemistry and much compound-building originates from this lively nature of the element.

In this way, the hydrogen in the universe has left its marks on planet Earth, mainly in the form of a variety of compounds. The most important of these hydrogen-containing compounds is *water*. Seen from outer space the liquid water on Earth, the gaseous/liquid water in the clouds and the solid water in the polar ice and ice-caps on land, all characterise the planet that casts blueish colour into the universe. Water in the clouds is an effective backcaster of light from the Sun, water in the oceans absorbs the solar radiation and the ice on Earth all reflect an appreciable part of the incoming light from the Sun. So how come that hydrogen ended up in these vast masses of water forms, vapour, liquid and ice?

It took various stages of evolution to form the water found on Earth. First, there was a need for having available all the oxygen atoms required for a globe full of H_2O molecules. That is a long and interesting story. Somehow, the first lifeforms on Earth interplayed with the energetic rays of the Sun and became small photochemical factories. One of the most common chemical cycles performed by these early lifeforms (like the primitive cyanobacteria) was photosynthesis, where the Sun is again playing a crucial role. Photosynthesis uses a captured photon from the sunlight to energise its process of catching carbon from the atmosphere. The excess oxygen atoms were made available by green algae and plants tuned into harnessing solar energy in the form of light and sequestering carbon.

Remember this term "sequestering" because it is going to be a theme through our story. The atmospheric oxygen was originally made by the cyanobacteria, and it really has to be treated as waste products of these microorganisms. Oxygen was gradually picked up by minerals on Earth resulting in oxidation. After a long time this oxidation was exhausted and free oxygen piled up in the atmosphere. So, in a few words, photosynthesis provides ample replenished oxygen and once present, oxygen stops the loss of water. The present ratio of oxygen in the atmosphere is 21 per cent, originally a result of a sustainable natural balance. The oxygen production also ensured that the water stocks on Earth were in a balance.

Both oxygen and carbon were originally created in the star factory. Light atoms like hydrogen and helium can easily escape from a small planet like Earth; they can simply float away because of their buoyancy and disappear into space. This happens when a leak of hydrogen occurs in volcanic eruptions and in geothermal areas.

With the advent of primitive lifeforms on Earth, the oxygen, carbon, hydrogen and a few other elements became key players in a new process: the formation of a biosphere. Here hydrogen enters a new stage as a building block of life. Plants use light to convert carbon dioxide and water into hydrocarbons. We know hydrocarbons as both a source of nutrition and fuel. In living organisms like our own bodies the hydrocarbons are converted by oxidation, a slow burning, into energy, water and carbon dioxide.

By mentioning carbon dioxide we are close to being able to take a breath and close this section. So we have used the Northern Lights to look 13.7 billion years into the past and have seen the key role of the lightest atom in the long history of our universe. Humankind in its present stage of development now has the opportunity to take over the process of the star factory and create its own fuel out of hydrogen.

In my car, which is slowly approaching Reykjavik, hydrocarbons stemming from ancient carbon dioxide captured by living organisms – deposited on Earth in vast oil-fields, have been recovered and burned in a combustion engine. The car makes a noisy approach to my home through the arctic night. I glance at the tank-level; it is getting much lower. With the cost of petrol this winter, it is, by the way, becoming far too expensive to run my car. Petrol is a hydrocarbon, offspring of a primitive plant life on Earth. How come that it is so expensive?

The point here is also about hydrogen because hydrocarbons are a primitive form of binding and storing hydrogen. The petrol in my car is a good example of the utilisation of the energy stored in hydrogen compounds by burning. The simplest way of harnessing the caloric value of hydrocarbons is to burn them: let them oxidise in the atmosphere and release heat. In this process the hydrocarbons unite with oxygen to form water vapour and carbon dioxide, CO_2. The water vapour creates moisture in the atmosphere. But there is a great amount of carbon dioxide which has a crucial effect on the atmosphere: it absorbs light of certain frequencies and does not let a hot Earth surface return the radiative heat into space. The carbon dioxide in the atmosphere works like a green-house roof. It keeps the rays of the Sun inside the atmosphere and causes heating. In a normal prehistoric Earth there seems to have been a balance in the carbon dioxide cycle of the ecosystem. But with the advent of the Industrial Revolution things got out of equilibrium. In this way, humans seem to be upsetting the natural balance of the sequestration of carbon dioxide. The resulting imbalance seems to be creating a vicious circle in the climate of the planet as its surface temperature appears to be increasing.

At the end of the last Ice Age, 12,000 years ago, the amount of CO_2 in the atmosphere was probably around 200 ppm, two hundred parts per million. Just before the Industrial Revolution, some 300 years ago, this level had risen to about 250 ppm. Today it is at 375 ppm. The increase of about 1.5per cent annually will quickly lead to a level close to 500 ppm near the middle of the 21st century.

One of the main questions we will study in this book is the challenge of reducing the use of the primitive hydrocarbon resources and bringing the greenhouse called Earth back to some equilibrium. Nature has chosen to bind together hydrogen and carbon in deposits. The proton is thus a very important building block of hydrocarbons. Humans understand the nature and physics of the proton and may have the possibility of utilising it in a systematic way. *Can we tame the proton in such a way that it becomes an important source of fuel* without linking it to carbon as in the case of hydrocarbons? Can we imagine a systematic creation of a new fuel-base founded on hydrogen without carbon? Will I be able to tell it to an interested reader without too much scientific formalism?

These and other questions go through my mind as I approach my house in Reykjavik with an ever-lowering petrol tank-level. The stray light of the city now blocks my view of the Northern Lights and the car works its way towards home. During my trip from the countryside I have spent a considerable amount of hydrocarbons and the atmosphere of Earth has received not only my breath of CO_2 but most importantly the "breath" of the car engine of CO_2. As a citizen of Iceland I am, on the average, responsible for some fourteen tons of man-made CO_2 every year. That is a large figure in great need to be looked into. Please join me tomorrow in further exploring the possibilities we have to make a change. Meanwhile, the Northern Lights continue to be seen dancing until the break of day.

Low on petrol

A nearly empty petrol tank calls for refuelling the next morning. The tank takes 70 litres. In Reykjavik it is easy to pick a station. My favourite station is the easily accessible Shell station in the eastern suburb of Reykjavik where Shell has been operating a hydrogen fuelling station since 2003 as a part of a demonstration project with three hydrogen fuel-cell buses. The petrol dispenser fills the tank in a couple of minutes.

The remarkable liquid called petrol originates from oil. Oil, in turn, originates from deposited living organisms that lived in basins on the edges of the world's oceans a long time ago. Over the long time, layers of organic matter have built up on the bottom. They contain among other elements: carbon, oxygen, nitrogen, phosphorus and of course hydrogen. In addition, the crust of the Earth is moving, rising or sinking and some layers can be buried in ever deeper rocks. If a layer of this oil-to-be is buried deeper than about two and a half kilometres in the sea bottom, the temperature and pressure reach conditions that favour a breakup of the organic molecules. Hydrocarbons are a major ingredient in some of these layers and the subsequent molecules. When the number of carbon atoms in the chains is from five to twenty, with the hydrogen atoms about two to four times more numerous, we have oil. Molecules with fewer than five carbon atoms are gaseous and the most common one is methane, CH_4. Oil experts call the depth of about two and a half kilometres the "oil window". Deeper burying of the oil will cause a further breaking up of the molecules and usually the magic depth of 2.5-5 km is the depth of the natural oil "factory" of Earth.

Oil, formed in this way, tends to be lifted upward; it floats on water and will migrate to the surface given the right conditions. Ancient civilisations on Earth have known oil or oil-like substances since antiquity. Bitumen, a hydrocarbon compound, was used for sealing baskets and the like. Baby Moses probably survived in a basket sealed with bitumen. Oil related substances were also used for medical purposes. When the Persians attacked Athens in 480 BC under the command of Xerxes, fires were initiated by arrows soaked in an oily substance.

The oil industry was started shortly after the discovery of the possibility of drilling for oil in the summer of 1859 in Titusville in Pennsylvania by Edwin L. Drake. Decades before an industry had been developed around a cumbersome production of a hydrocarbon, kerosene, from coal. Soon after Drake's discovery, oil refineries were making oil for the newly invented combustion engine, first developed by Nikolaus Otto in Germany 1861. About simultaneously with the Drake well, an oil area in Baku, Azerbaijan became known, and in the four last decades of the nineteenth century the oil industry and the combustion engine became symbols of human progress.

There was an exponential growth of oil discoveries and use.

The advent of the automobile around the beginning of the twentieth century finally created an explosive demand for oil. Furthermore, it should be noted that oil is an extremely important basis for a whole industry related to lubricants and polymer production and resulting in all sorts of, for example, plastics. Polymer engineering is a key area in modern materials science and without oil as a source the industrial world would be totally different.

In the heydays of the oil-boom, the feeling was that this was a gift from heaven. Today, over fifty thousand oilfields have been discovered and the gross production has exceeded 25 billion barrels of oil annually, or some 3.4 billion tons. But was there a limit for the amount of this easy energy? In the beginning there were no such worries. The answer is of course yes, given the fact that the Earth has a finite size. So, unless the process of oil formation is faster than its use, oil is bound to be limited. Again here, the Shell oil company enters the story.

Over the years, some remarkable work has been done in the Shell geosciences headquarters in USA. In the 1950s a prominent Shell geophysicist, M. King Hubbert published a study of the estimated development of the United States oil production for the nearest future. The Hubbert methodology was based on results similar to those used by biological population analysis. In Hubbert's analysis, the U.S. oil production would reach a maximum in 1970 and then gradually decline thereafter. The scientific community did not accept this news so easily and there was considerable controversy surrounding the subject for at least two decades. But looking back now, one observes that in fact the U.S. oil production culminated in 1970 with about 11 million barrels per day and is currently only 2/3 of what it was at its peak!

Modern assessment analyses of oil deposits use Hubbert's method as a standard. The implications seem to be that the oil deposits on Earth are certainly limited (fig.3). Discovery of new oil fields seems to be slowing and the quality of new discoveries is of a lower grade, often requiring much more elaborate refinery or processing methods with subsequent pollution problems. In 2007 the largest single country in oil production was Saudi Arabia with about 10.7 million barrels per day. In their wake, Russia is following quite close with about 9.7 million barrels per day, and the United States is in the third place with just under 8.4 million barrels per day or just under a third of the production over three decades ago. The notorious hurricane Katrina in late 2005 harmed some of the important oil rigs and refineries in the Louisiana area with the effect of further slowing down the production rate for at least a limited period.

The distinct difference in the production profile of the largest producers is their varying number of producing wells. The giant Saudi Arabia extracts its oil from just under 1,600 wells. Russia, responding to higher oil prices at the turn of the millennium,

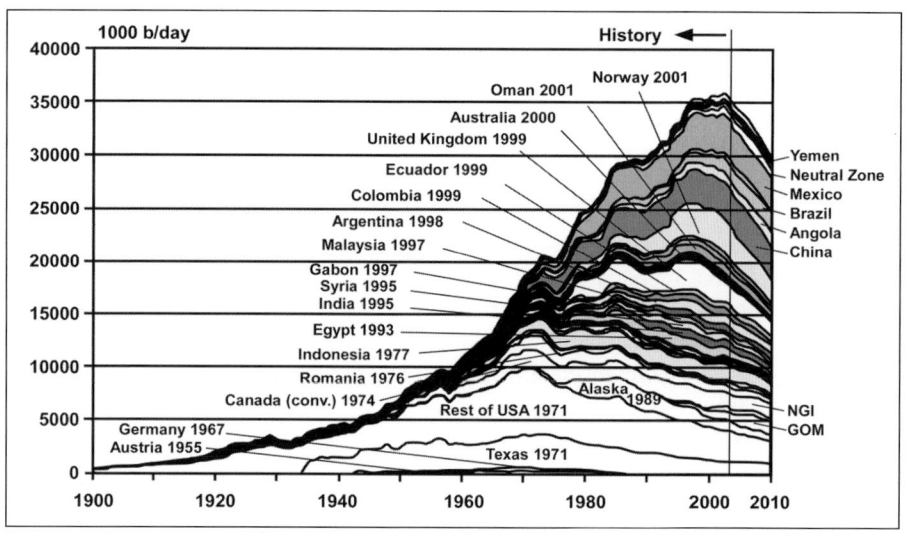

Figure 3. Oil production of countries outside OPEC and the former Soviet Union as predicted by a number of sources.

now produces from about 41,000 wells. Remarkably and quite significantly, the U.S. pulls its oil out of over half a million wells.

Early twentieth-century military build-up in Europe created a new dimension to the importance of oil and its applications. Winston Churchill is famous for taking the step for Britain to convert Her Majesty's Fleet from coal to oil. There has been a direct relationship between oil production and oil shortage and the price of oil. During the Second World War, Nazi Germany, partly cut off from oil delivery, had to make use of a method for making oil from coal and steam in the so-called Fischer-Tropsch process, which we will learn more about later. South Africa developed this method further in the shortage situation that resulted from an international export ban to the country.

In 1973, the world saw the first major repercussions of fossil-fuel insecurity when an oil crisis threatened the financial stability of the industrial countries. The Gulf war scenario in 1991 was greatly affected by the presence of oil and burning of oil fields became a symbol of the vast destructive power of that armed conflict. Price volatility of oil is a fact. Oscillations in oil and gas prices can be enormous and result in a very shaky foundation for basing economic forecasts, pricing of goods etc. In August 2005, the time around the hurricane Katrina, the oil prices rose to an all-time high of $70.85 per barrel on the New York Mercantile Exchange, rising from a low $47 in May. In 2006 the record high of $78.40 was set on July 13th from which date they lowered appreciably. Initially the price then lowered somewhat before re-stabilising approaching the $100 feared by economists and politicians in the autumn of 2007.

The projected increment in the world oil production is expected to grow from 78 million barrels per day in 2002 to about 119 million barrels per day in 2025. A new player on the scene, the emerging economy of China, is expected to grow in demand for oil by a whole 7.5 per cent annually until 2010 and somewhat lower in the 2010-25 period which explains why there are "only" 119 million barrels in 2025. Members of the Organization of the Petroleum Exporting Countries (OPEC) are expected to supply the majority of the produced oil in the period up to 2025. The non-OPEC countries are expected to supply an extra 17 million barrels per day in the period up to the same year.

The view taken in this book will not be extreme. We do not intend to set a certain number of decades left for oil to be a major fuel on Earth. We will even expect oil to be with us for a long time. Some countries will have access to oil for all up to a century. Some countries may be forced to go into pristine areas for oil drilling and using ever more sophisticated methods to extract oil out of difficult geological formations, tar sands and oil shales. Oil will probably be extracted to the extent when the last drop is squeezed out of the last rock in some places. It is generally recognized that oil is a source of insecurity for the world at large. Oilfields are mostly owned by governments and will continue to be a destabilising factor in world affairs. All we will say at this point is that oil seems to be a limited gift to humankind and may form a short period in its history and that much innovative thinking will be needed in order to secure a new basis for the fuels of the future. In the meantime, the importance of another aspect of the oil story will become increasingly prominent. This is the effect of oil and hydrocarbons on the global climate scenario which will be discussed in the next chapter.

Exhaustion

When the petrol, or for that matter diesel, is utilised in a combustion engine, the hydrocarbons are burned in ambient air to form mainly water vapour and CO_2.

Much of the energy released in the burning of petrol is converted into heat. Only a fifth to a sixth of the energy released goes into turning the wheels of our cars. This ratio of useful and available energy is usually referred to as the *efficiency* of a combustion engine and detailed calculations are available for what is called well-to-wheel or tank-to-wheel efficiency of a motorcar.

What happens to the water vapour and the CO_2 from the engine burning process?

Well, the water vapour is returned to the atmosphere and thereby joins the water contained in Earth's ecosystem. Some of it evaporates, falls as precipitation, is returned to seawater or forms clouds floating in the sky. When the rays of the sun hit Earth, some of them are reflected by the clouds, others are absorbed in the oceans and lakes. A part of the energy is also absorbed by the water molecules and in that sense trapped in the atmosphere. It has been estimated that without the cloud cover, the average temperature on the surface of Earth would be about 4 degrees higher.

The physics of CO_2 and H_2O is in many ways very interesting. We have to imagine a CO_2 molecule composed of one carbon atom between two oxygen atoms. H_2O is a similar three-atom molecule. The forces keeping the atoms in the molecule are "soft" in some sense; they result in a gentle vibration of the molecule like in spring pendulums. Photons can excite the movement of the molecular vibrations and effectively be converted into molecular vibration energy. An appreciable part of the infrared radiation that is emitted from the heated surface of the Earth, is not capable of escaping out into space because of the trapping effect of these vibrating molecules. This is known as the green-house effect. CO_2 as well as H_2O are very important green-house gases. Some other important gases in the atmosphere are also green house gases. Methane, CH_4, ozone, O_3, nitrous oxides, NOx, and chlorofluorocarbons are all in that category. Because the atmosphere is such a good absorber of longwave infrared radiation, it effectively forms a one-way blanket over the Earth's surface. Visible and near-visible radiation from the Sun easily gets through, but thermal radiation from the surface can't easily get back out. In response, the planet's surface warms up.

So what characterises the natural cycle of all the burning of carbon containing materials? It seems that in the course of time a lot of natural systems on Earth have been working together to harmonise the CO_2 cycle. The trapping of carbon in hydrocarbons was mentioned earlier. Photosynthesis in forests around the world binds CO_2 in huge quantities. Rocks can absorb the molecule CO_2, precipitation absorbs it, binds it to minerals in rocks and finally returns it to the ocean where it will be either stored in solution or precipitated in sediments.

In the middle of the Pacific Ocean, on the islands of Hawaii, CO_2 in the atmosphere has been measured in a station situated near the top of Mauna Loa Mountain since 1958. These data constitute the longest continuous record of atmospheric CO_2 concentrations available in the world.(See figure 4.)

Measurements in ice caps and sedimentary deposits have revealed that the amount of CO_2 in the atmosphere around the end of the Ice Age must have been close to 200 ppm, two hundred parts per million. Before the industrial age and the subsequent coal and oil exploration the amount of CO_2 in the atmosphere was about 275 parts per million. It equals that in a sample of one million atmospheric molecules, about 275 mol-

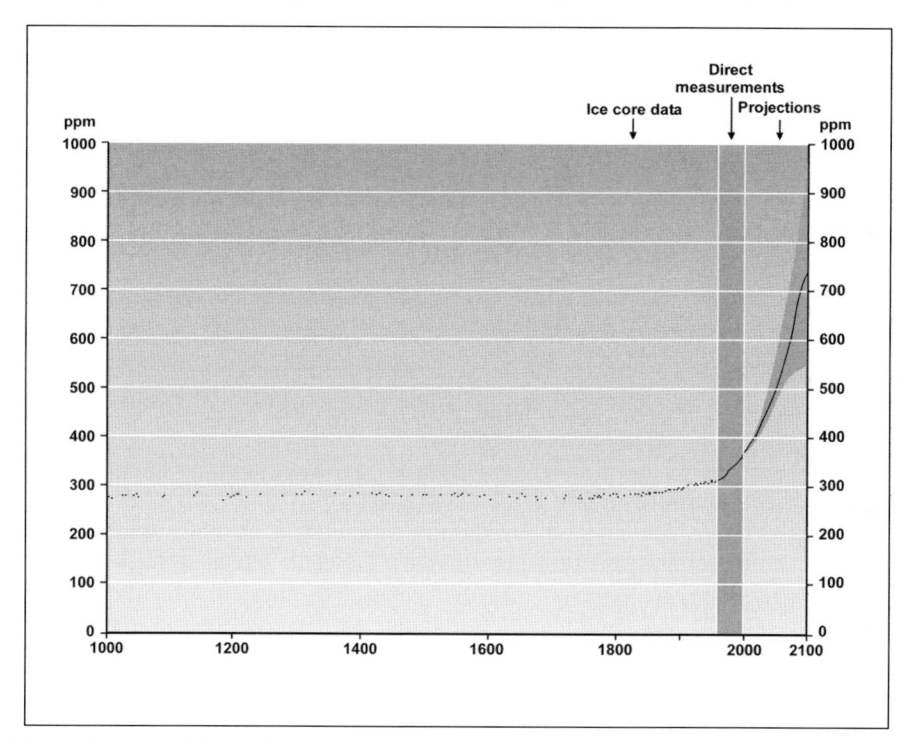

Figure 4. Past and future CO_2 atmospheric concentrations as shown by the Intergovermental Panel on Climate Change, IPCC.

ecules are carbon dioxide. The measurements on Mauna Loa indicate at present a 370 ppm level. The measurements additionally show a seasonal effect; it seems that summers in the Northern hemisphere lower the value of CO_2 probably by plant uptake; winters increase the carbon content once more, as increased need for oil and gas for heating shows up with a net increase in the atmospheric CO_2 proportions.

A careful reader may at this point protest and ask why it is concluded that the man made-anthropogenic- effect is causing this and what proportion would be attributed to the expected natural oscillations and cycles. This is a valid question with which we will pause for the moment.

The measurements from ice caps and sediments, usually termed a part of paleoclimatology, reveal that in the course of Earth's history much change has taken place. It seems that in periods of vegetation, volcanic eruptions, precipitative spells or soil formation eras, CO_2 was changing and oscillating quite dramatically. There is a correlation between CO_2 proportion and surface temperature (Fig.5). There are a number of accepted natural fluctuations in the data. But one thing seems to be undisputed: No species and no processes have been systematically disturbing the balance of CO_2 absorption as we humans have done through the drilling, exploration and consequent burning of oil and the massive burning of coal. Whatever can be expected from natural fluctuations, it seems now without doubt that the role of anthropogenic CO_2 is considerable and scary.

The sceptic may even question the validity of claims that climate is being affected by CO_2 increases. To shed light upon that aspect we could go to totally different sources and for example look at insurance claims from the aftermath of catastrophic weather in the world, storms and hurricanes, as has been studied by the Worldwatch Institute in Washington – and their answer will point in the same direction. In each decade of the past century the insurance claims in the U.S. from damages of hurricanes showed a fast rising tendency. In the eighties, the insurance claims doubled from the previous decade; and in the nineties a tripling took place. A very thorough analysis by Evan Mills in *Science* in 2005 calculates and renormalizes with inflation, population increase as well as insurance penetration and concludes a very significant doubling in two decades. Meteorologists assign a given total energy to a hurricane. This index is called Accumulated Cyclone Energy (ACE). The hurricane season of 2005 was particularly active and the 2006 season seems considerably less energetic. Figure 6 shows the inflation-adjusted global costs of extreme weather.

Looking for wider implications, what about the effects on biology? Paul Epstein, of the Center for Health and the Global Environment at Harvard Medical School, has pointed out that extreme weather events reflect massive and on-going changes in the climate on Earth to which biologic systems on all continents are reacting. For clarification, he points out how mosquitoes which can carry many diseases, are very sensitive to temperature changes and how warming of the environment boosts their rates of reproduction, and the number of blood meals they can take as well as prolonging their breeding season. Furthermore this shortens the maturation period for the microbes they disperse.

Changing climate affects fungal diseases in crops, coral reef growth and health of

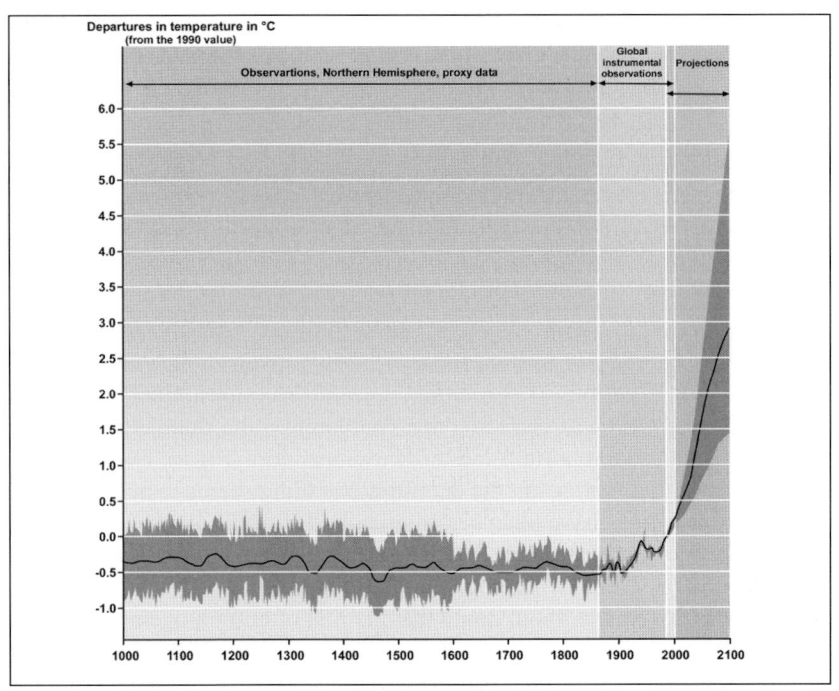

Figure 5. Variations of the Earth's surface temperature: year 1000 to 2100. Based on IPCC.

forests to name but a few of the sufferers. In his article in *The New England Journal of Medicine* Epstein lists a number of complications ranging far and deep into a variety of biological systems and their connecting ecology.

Also, there would be a reasonable claim to look for effects going in the opposite

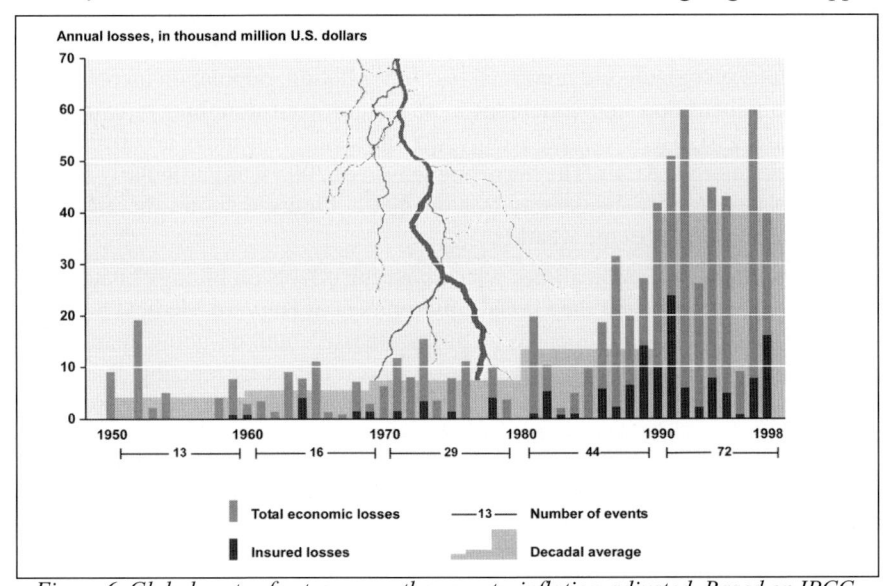

Figure 6. Global costs of extreme weather events, inflation-adjusted. Based on IPCC.

direction. It is for example well known that cosmic radiation can enhance the production of aerosols, small particles that are the seeds of cloud formation on Earth. Recent studies by the Danish Space Research Agency have shown that due to decreased cosmic radiation on Earth over the past century, low and sun reflecting clouds have been more scarce. This would have a cooling effect.

Another of these cooling effects could well occur in the North Atlantic by a slowing down of the Gulf Stream due to disturbances in ice melting off Greenland and Canada. Other similar effects would be associated with the concept of climate cooling. It is well known that the burning of oil and coal will result in soot and particulate air pollution which in turn will act as an increasing reflecting power of the atmosphere when hit by sunlight. The same will hold for huge volcanic eruptions. The short term consequences will be slight cooling and temporary blocking of sunshine. We remember the calculations made by experts visualising the aftermath of a nuclear war on Earth with subsequent effect on the cloud cover and a phenomenon termed "nuclear winter". History shows an often stunning relationship between volcanic activity on Earth and dramatic climatic and social implications.

In Iceland historical annals disclose that an enormous volcanic activity took place during the Lakagigar eruption which began in early June 1783. By midsummer a black mist is said to have hung over Europe, reaching from Finland to the Balkans.

The European press reported "blood-red morning sun" with a midday light so weak that the Sun could be viewed with the naked eye. By July the mist had reached China and Japan. In Iceland the eruption led to the subsequent famine and population reduction during the cold winters that followed what is seen as the largest ever lava production in historical times.

The French Revolution in 1789 marked a turning point in the evolution of republicanism and civil rights. One daring theory even suggests that the French Revolution partly had its roots in the Lakagigar eruption, when, as previously mentioned, the effects of the mists had altered weather patterns across Europe with accompanying crop failure and public unrest.

Large eruptions continue to take place in Iceland. Figure 7 shows activity in the Krafla volcano in Northern Iceland which had a fissure eruption in 1984.

It has been pointed out that if no action is taken in limiting the emissions of CO_2 in the near future, its atmospheric proportion will rise to about 550 ppm before 2050. In the next chapter we will address the possibilities open to the energy system on Earth to attempt to slow down this tendency. Thus it is very important when trying to find new means of providing energy to address CO_2 emissions and other environmental effects of energy use. Especially since the Earth's climate and environment is so undeniably out of balance.

A report commissioned by the British Government and written by Sir Nicholas Stern in 2006 has been seen as a grave reminder of the effect of climate change. Sir Stern has warned that the effect of global warming can create economical disruption and lead to economical consequences similar to that of the World Wars and the Great Depression in the former half of the 20th century.

The report urges Europe to limit the greenhouse gas emissions to about 30 per cent

Figure 7. The Krafla Volcano erupting in 1984.

before 2020 and 60 per cent before 2050. Never before has a report like this been commissioned by a leading Government. The assumption of the report is that combating climate change will cost about one per cent of the GDP.

In January 2007 The Intergovernmental Panel on Climate Change published the results of years of studies of several thousands of climate experts where it was finally confirmed that a large part of the climate change observed was caused by anthropogenic effects. In October 2007 the panel, together with former US Vice President Al Gore, was awarded the Nobel Peace Prize for their work on environmental affairs.

We will now spend some time studying the various energy sources and systems used on Earth, before further discussing ways to reduce CO_2 emissions through decarbonisation.

Burning and converting

Thermodynamics: the flow from hot to cold

Energy can not be created or destroyed. It can be transferred or converted from one form to the other. The first law of thermodynamics concerns the conservation of energy. The engine in our automobile converts heat from the burning of petrol into kinetic energy of motion.

During the conversion from one state to the other, the second law of thermodynamics states that the final state will have less potential energy than the first; some energy will be lost as the world has a tendency to maximise its so-called entropy.

In the course of civilisation many machines have been developed to assist or extend the muscular power of humans or domestic animals.

Waterwheels were used to power grinders of grain and pump water; Windmills were utilised for the same purpose and mechanical looms were applied for weaving. Both technologies harnessed the natural flow of water or wind. Converting heat energy of a fuel to rotational energy of a machine was a major achievement in the development of society.

Steam engines were the first major engines of the Industrial Revolution. James Watt, the young Scottish instrument maker, was able to develop a steam engine that worked better than any tested before. The year was 1765. This engine is an example of the way to put heat into work. A boiler heats up water into steam. The higher the pressure of the steam, the higher the energy content, and in a given fixed volume of steam, increased heat input will lead to higher pressure.

The steam pressure is ideal for pushing a piston, which in turn is linked to a rotating axle. A massive flywheel is used for stabilising the power and make sure the piston returns from a fully extended posture back into a position ready to receive more steam and complete a full cycle.

This is the place where we introduce the concept of efficiency. One of the laws of thermodynamics is that work can be put into heat; but heat can not be converted into work with a full 100 per cent efficiency. There will always be losses. The heat content in my petrol tank will, in fact, only to about a fifth be turned into the driving power of the wheels. The remaining 4/5 are returned in the form of heat and will be of no use, merely heat up the universe!

There are numerous ways of converting heat into work in a way related to the steam piston. Figure 8 shows the principle. They are all commonly called heat engines. The internal combustion engine we know from our cars is an example of a heat engine. The revolutionary engine – the first to burn fuel directly in a piston chamber – was developed by the German engineer Nikolaus Otto in the 1860s and 70s. Today, the four stroke cycle is often referred to as the "Otto Cycle" in his honour.

Beyond combustion engines, there are turbines which are the most common devices for converting heat of oil or hydrocarbons into electricity. For this purpose an electric generator is connected to the rotary shaft. In an electric dynamo, a large bundle of carefully woven copper wire is made to rotate in a strong magnetic field. As a result, electric current is induced in the wire of the dynamo. The electricity can be hooked to a network to power the electric machines all around us. The disadvantage with this method of producing electricity is the emitted CO_2 from the primary source. Electricity production from fossil fuels is probably the largest single source of CO_2 emissions on the planet.

The overall efficiency estimate of a heat engine was first described in 1829 by a young Frenchman, Sadi Carnot. The main conclusion of Carnot was that a heat engine could never deliver more efficiency than that which was limited by its own temperature of heat source and the limits set by the temperature of the cold sink involved. Heat engines take energy from a hot reservoir, convert a part of it to mechanical energy like the rotation of a crankshaft, and return the "working fluid" to a colder reservoir, the sink. The second law of thermodynamics states that the thermal efficiency has an upper limit and can never reach 100 per cent.

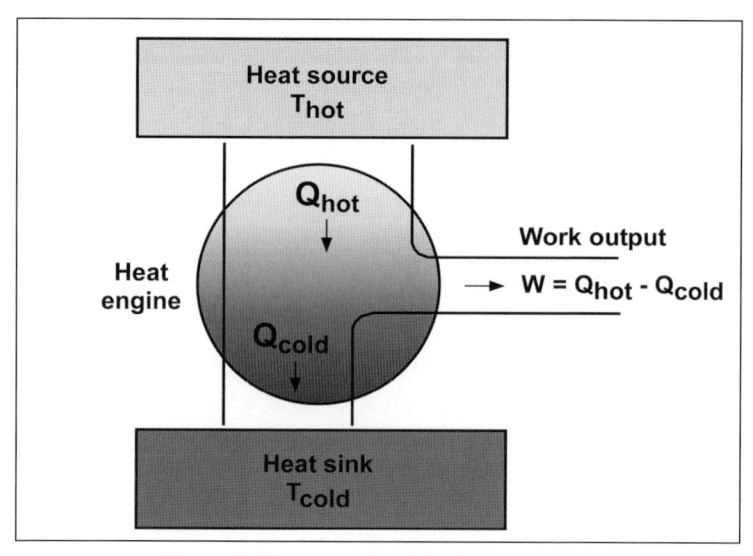

```
                    Heat source
                      T_hot

                       Q_hot
                         ↓              Work output
            Heat                    →  W = Q_hot - Q_cold
           engine
                       Q_cold
                         ↓

                    Heat sink
                      T_cold
```

Figure 8. The principle of the heat engine.

Imagine a steam turbine at 280°C which is cooled with a 20°C cold source. In Carnot´s theory the maximum efficiency would be equal to the difference in temperatures of these opposite sources divided by the temperature of the hot reservoir. In a calculation we would have to express the temperature in the units of Kelvin. In this way we can quickly see that a 100 per cent efficiency is probably unattainable unless the cold source is close to absolute zero temperature on the Kelvin scale. This theory was quickly termed as Carnot´s maximum efficiency. The process of fossil burning in combustion related engines is a part of the Carnot era in energy conversion. It is characterised by huge losses and inefficiencies. We will return to these thoughts later when we explore the taming of the proton and the so called free energy in more detail in a chapter on fuel cells.

There is finally one conversion aspect to be added at this stage when we study general energy conversion and efficiency. This is the concept of hybrid utilisation in transport vehicles. The hybrid concept is ideal for cars with an electric drivetrain. One has to bear in mind that in a classical automobile a large part of the kinetic energy achieved when moving fast along the road is converted into heat as soon as the brakes are used to bring the speed down. The kinetic energy increases with the mass of the car and doubling the speed results in a fourfold kinetic energy. The brakes heat up when braking, thus converting kinetic energy to heat. This heating effect is often visible on a racecourse!

By utilising an electromagnetic braking system the hybrid car converts the mechanical energy into electrical energy and stores this energy aboard by batteries and sometimes supercapacitors. The ideal hybrid car has an electromagnetic brake on all wheels. The famous Toyota Prius, and other hybrid cars by a number of manufacturers, are an ode to this concept and deserve much applause for innovative and environmentally responsible thinking.

In the next sections we will go systematically through the available resources of primary energy on Earth and discuss some of their limitations and possibilities. We will divide them into renewable and non-renewable sources. For energy to be defined renewable it has to be derived from resources that are regenerative or for all practical purposes cannot be depleted.

Non-renewable types of available energy sources

We will now go through most known energy sources used in the world today. We will attempt to assess their availability and capacity to fulfil the expected needs as well as the anticipated development of their utilisation in the future. Figure 9 shows the relative importance of various energy sources in the world and their projected values.

We started with oil in the last section and will now continue with: *Natural gas*

Wherever oil is harvested, as we discussed before, it will be accompanied by different amounts of the lighter hydrocarbons, in the form of natural gas. Right now, natural gas is believed to be the fastest growing component of the world's primary energy consumption in the years to come. The annual rise in the natural gas consumption is expected to be around 2.3 per cent compared with the previously mentioned 1.9 per cent for oil. About two-thirds of the increase in gas demand is in the industrial and power generation sectors, while the remaining one-third is in space heating of buildings and homes.

The fraction of natural gas in the energy portfolio of the world is about 23 per cent and will rise to about 28 per cent before 2025 according to a U.S. Department of Energy estimate. Most of the rise in the use of natural gas comes from the electricity production sector which accounts for about half of the demand. Increasingly, natural gas is expected to replace oil and coal based electricity generation plants with the resulting net reduction in CO_2 emissions per unit electric energy as we will discuss in a later section on decarbonisation.

Furthermore, natural gas will provide an interesting lower emission alternative for automobiles, and advances in gas engine technology and gas storage will help to increase the market share of natural gas for decades to come.

It is difficult to use Hubbert's methodology of assessing natural gas reserves. *The Oil and Gas Journal* has estimated that most of the world's reserves are located in the Middle East or some 34 per cent of the world total, and Europe and the former U.S.S.R. with 42 per cent of total world reserves. The United States, by this calculation, possesses three per cent of the world's total natural gas reserves.

Coal

The Chinese were probably the first to mine coal on a large scale. A coal mine is said to have been built in Yexian County in what today is a part of the Henan Province in 210 AD. Marco Polo mentioned in his travel accounts after returning to Italy in 1295 that the Chinese used for fuel a kind of "black stone" which "burned as easily as wood but had much stronger flame and could last until the next day." In the 13th century a charter dealing with and recognising the importance of coal supplies was granted to the freemen of the craft guilds of Newcastle, allowing them to dig for coals unhindered.

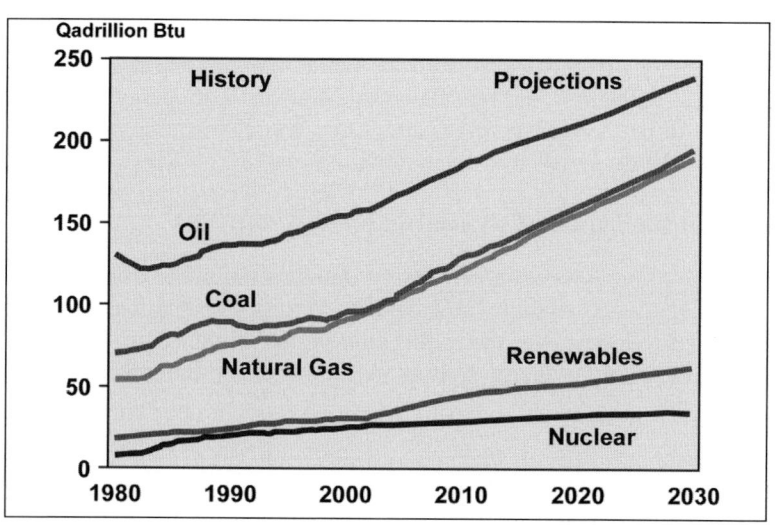

Figure 9. The relative amount of energy in the world and the projected use. The unit of Quadrillion Btu or Quad refers to a very large quantity of British thermal units. The total annual energy consumption of humankind amounts to some 440 Quads. Source: Energy Information Administration.

When James Watt powered his steam engine, coal was the most important source. Coal became a crucial part of the steelworks of the Industrial Revolution where iron ore, limestone and coal were fed onto the top of what was termed blast furnace with ambient air being blown through the bottom. Useful Iron and less precious slag came out of the furnace. An ingenious construction industry was at that time based on all sorts of slag and coal treatment or coke by-products.

The coal deposits on Earth are quite enormous. The coal usage on Earth is increasing by 2.5 per cent annually. Presently it accounts for about a quarter of all energy used in the world similar to natural gas. Of the coal shipments worldwide, about 2/3 go to electricity production and 1/3 goes to industrial uses. Unfortunately, the coal thus used on Earth ends mostly in the atmosphere as CO_2. Lower grade coal deposits also increase even further the content of sulphur in the emissions. Coal also contains many trace elements, including arsenic, thorium and lower levels of uranium and other naturally-occurring radioactive isotopes. While these substances are trace impurities, if a great deal of coal is burned, significant amounts of these substances are released.

There is plenty of coal in the world. The International Energy Agency has estimated that economically recoverable coal reserves on the planet are up to one trillion tons, enough to satisfy another two centuries of coal burning. Only ecological reasoning will limit coal use in the future.

China which has huge deposits of coal and a dominant position in steel making is expected to continue to use coal long into the next century. The Energy Information Administration, in its International Energy Outlook 2005 envisages coal and natural gas as accounting for nearly 2/3 of the electricity production in the world in 2025.

Nuclear energy

Before turning our attention to renewable energy, it is important to dwell on another energy form on Earth: nuclear fission. In the grand star factory we mentioned earlier, when stars exploded from gravitational collapses, a vast number of radioactive elements were produced. These elements are radioactive because of an inherent instability in their nuclear composition. On Earth there are deposits of such elements and compounds which can be utilised as a main source of thermal energy.

Uranium is an element that can be found in appreciable amounts in various areas of the globe. Most uranium has a very heavy nucleus which is about 238 times heavier than hydrogen. When a neutron, the uncharged nuclear particle, is made to strike a uranium nucleus (the isotope U^{235} which contains a fewer neutrons than the heavy one) it is split into a variety of fission products. One important path leads to, for example, tellurium and zirconium plus a couple of neutrons or, alternatively, it splits into barium and krypton plus three neutrons. According to Einstein's relationship between mass and energy, the difference in total mass before and after the fission, is converted to energy in accordance with the $E = mc^2$ relation which he put forward in his "annus mireabilis" in 1905. Here m is the mass difference and c is the speed of light which has to be squared to get the result.

Nuclear reactors use a "fuel", namely a fissionable substance such as uranium-235. To do this the uranium from ore is enriched to about three per cent uranium-235 content and then used in the form of UO_2 pellets. Rods, containing materials, such as cadmium or boron, are used to control the fission process by absorbing neutrons and calming down the process. In this way, the reaction chain is kept self-sustaining and the reactor is prevented from overheating.

In most nuclear reactors the heat produced by the nuclear reaction is carried by a fluid such as water or liquid sodium to a heat exchanger in which steam is generated. The heat exchange liquid moves in a closed loop and the steam so generated is used to drive an electric generator.

Nuclear fission is a very important energy source on Earth and some countries are quite dependent thereupon. France, for example, receives a considerable proportion of its electricity from nuclear power. In France the art of nuclear power for peaceful purposes has reached a high degree of maturity.

In the present energy portfolio of Earth's population, nuclear power accounts for about 6.2 per cent of the total energy production. The consumption of electricity generated from nuclear power was about 2,560 billion kilowatt-hours in 2002 and is expected to rise to 3,270 billion kWh in 2025.

The International Atomic Energy Agency (IAEA) has estimated the economically useful uranium reserves on Earth to be 4.7 million tons with 54 per cent to be found in developing countries and three industrialised countries, Australia, Canada and the United States holding the largest deposits. Just over a couple of million metric tons of uranium is found in recoverable vein deposits; whereas a hundred thousand times more uranium is found in very dilute amounts, like, for example, in seawater. Radioactive waste is the main Achilles' heel of nuclear fission. The fission products with the longest half-life are for example plutonium 239 with half-life of 24 thousand years. The term "half-life" refers to the time it takes to reduce the radiation level to half of its initial value.

The most effective way of storing radioactive waste is in halite (salt) or anhydrite (calcium sulfate) beds. Within a lifetime of a person in a nuclear-based society the typical amount of radioactive leftover for one individual corresponds to about the size of a tennis ball. Environmentalists do not accept current radioactive waste disposal technologies and a great dilemma exists about the usefulness of nuclear fission energy.

Many things can be said about nuclear fission. The problem of long lifetime of the waste products is a huge hurdle. But the positive aspect of nuclear fission concerns the negligible level of CO_2 emissions with the exception of CO_2 produced in the life cycle of plants, construction and so on. The nuclear industry is very interested in using nuclear energy to make a fuel, hydrogen, which by itself is renewable in nature. We will come back to this when we discuss infrastructures.

As a physicist I have always been fascinated by one aspect of nuclear energy. This is the energy of fusion which we studied in the discussion of the star factory. In the case of fusion, two light atoms fuse into a heavier one. In order to create fusion in a laboratory, one has to imitate the process of the Sun: to bring together in a frontal collision two deuterium atoms which in turn will be fused into a helium atom! Such a collision is pretty difficult to manage and control, especially because the temperature of this man made "sun" is very high and no materials known have the ability to hold such a hot soup in a controlled manner. This is why experimental fusion reactors contain so-called magnetic mirrors, intense magnetic fields that can be configured in space in order to hold the soup in a virtual magnetic "bowl". When I was a student at the Cavendish Laboratory in Cambridge, UK around 1980 there were exciting news about what was named cold fusion where some scientists thought they were seeing a fusion of hydrogen atoms at low temperatures. The most obvious explanation rested on the assumption that the so-called quantum mechanical tunnelling of two hydrogen atoms in the solid phase that was expected to be the effective fusion that would lead to the formation of helium. Unfortunately, the findings of the cold fusion experiments were found to be in serious error because of a flaw in calorimetry, the measurement of the temperature rise of the sample. Cold fusion proved to be full of confusion!

I am convinced that the future will see the power of the Sun harnessed directly in small "suns" on Earth. My guess is that it will start beyond 2050, when the anthropogenic CO_2 content of the atmosphere will have reached a double of its value from present (assuming no political moves in the opposite direction). In order to make fusion possible, advances in materials science and plasma technology will have to be made. My feeling is that most of the ingredients are there and perhaps a generation of scientists away from realisation. The only remaining issue will be how to convert nuclear energy into a transportable fuel. There seem to be two intertwined solutions: electricity from chemical storage batteries and/or hydrogen storage and fuel cells.

Renewable energy sources

Photovoltaics

At this point there is the following observation to be made: The Sun can in many ways be seen as a source of a variety of renewable energy sources. By renewable we then mean energy sources that can be assumed to stem from a long standing, almost unlimited source. Of course our Sun has a limited lifetime, but we will treat it as eternal. After all, the Sun will hopefully continue to shine on Earth for millions of years to come. The Earth receives about the equivalent of 1,372 Watts of solar radiation for each square meter of atmosphere facing the Sun. Most of this energy is either reflected into space or absorbed through collisions with molecules in the atmosphere or from the planet´s surface. As a result only a third of this power is available on average.

But this solar origin of renewable energy also creates what we could call solar energy derivatives. By this we usually mean wind, ocean currents and hydroelectric energy. Also the movement of the Moon around Earth creates forces of tides which in turn are a source of renewable energy. Finally, ocean currents, often a product of a combined effect of Sun heat and planetary motion, are a true renewable energy source. Most of the energy we use on Earth is produced from burning hydrocarbons which can be seen as solid sunshine.

The Frenchman, Edmund Becquerel, observed already in 1839 a relationship between light and electricity and of the photovoltaic nature of matter. In 1905, Einstein described the nature of light and the photoelectric effect on which photovoltaic technology is based, for which he later won a Nobel prize in physics. The first photovoltaic module was built by Bell Laboratories in the United States in 1954. Photovoltaic effect happens when a light photon kicks an electron out of its place in a conducting or semiconducting solid and is a very important potential source of electricity.

Solar panels, made of silicon or related semiconducting compounds, are an increasingly important source of electricity on Earth. They are usually referred to as Photovoltaics or PVs. Improvements in the technology have produced maximum efficiencies of up to 20 percent in commercially available PV units, and this technology is steadily improving. Current wafer based crystalline silicon cells are expected to reach up to 25 percent efficiency. CO_2 emissions of PVs have to be calculated from life cycle analysis mentioned earlier. In order to assess the CO_2 associated with PVs, one has to calculate the burning of hydrocarbons or whatever in the process of making silicon for the wafers used in the panels. On the whole, CO_2 emissions through a total life cycle of a solar panel are much lower than from a corresponding source in the old hydrocarbon economy.

Solar PV production in the world passed the 500 MW production capacity level in 2002 and had increased tenfold in the previous decade. Today´s installed capacity exceeds 1,000 MW (1GW), or about the size of a nuclear power plant. At present, some ten producers dominate with about 90 per cent of the world market. It is difficult to quote prices for installing PVs. Remote PV systems of up to 1 kW in size can vary from 10-18 USD per Watt and with almost the same amount spent on high reliability batteries and control units. Payback time can be as low as two to five years. Long lifetimes help to make this investment sensible. Presently, the lowest generation expenses one hears of are in sunny areas with low investment costs, where the cost per kWh can fall below 0.20 USD. One of the exciting developments of this technology involves the merger of architecture and PVs where the latter become building materials with a double function.

The installed PV power in the world is expected to reach 11 GW in 2010.

Concentrating solar power

Solar harnessing can also be done in such a way as to recover both the electric part of the conversion as well as the heat part. In this way there are on the market solar panels generating both heat and electricity. We say that the total efficiency in these combined power plants is much higher than in conventional electricity-only production.

Let us take a look at the direct concentration of solar power (CSP). In France around 1860 there were solar powered steam engines constructed by the inventor Auguste Mouchout for several applications. In 1900 a 45 kW sun tracking parabolic sun-catcher was installed by the American Frank Schuman in Meadi in Egypt with an accompanying solar motor. The most impressive demonstration of this technology was done with the nine generating stations in the Mojave Desert in California in the last decades of the twentieth century.

Most CSPs consist of four basic units: concentrator, receiver, transport-storage and power conversion. There are mostly three types of systems in use as shown in fig. 10. The so called *Trough System*, commonly known as "solar farms" uses linear parabolic mirrors to reflect sunlight. A second type is the *parabolic dish system* which col-

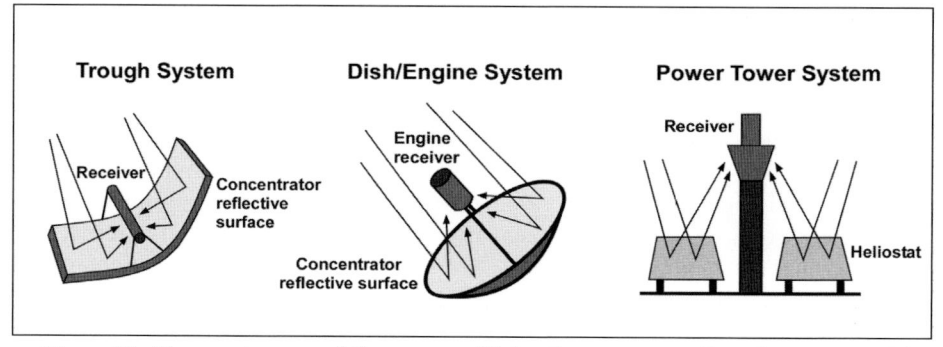

Figure 10. Three main types of Concentrated Solar Power systems. Trough system, Dish/Engine system and Power Tower System.

lects sunlight through a parabolic solar-collector. Finally, there is the *Power Tower System* which consists of large sun-tracking mirrors, called heliostats, which can concentrate solar power onto a central tower-mounted receiver. At the receiver, the ray energy is used to produce steam for a turbine generator. I visited a plant outside Sevilla in Spain where an 11 MW power system is being built based on a solar farm and a power tower with heliostats which look like a modern day "Stonehenge". In Andalusia a 3x50 MW electric CSP is being built.

CSP technology is currently believed to be able to provide energy at the cost of between 0.10 and 0.15 USD per kWh. The industry is working hard on improving materials and technology aiming for about 0.03 USD per kWh before 2025.

The installed capacity world-wide around the turn of the millennium was about 370 MW electricity with an output of 1 TWh annually. Industry estimates are around 14-20 GW expected to be installed in 2020.

Wind energy

Wind energy is the result of the moving masses of air around the globe and is a derivative of solar power with some effect from the turning of the Earth. The unequal distribution of heat around the planet, the presence of mountain plains or oceans all affect the wind and weather machine. Harnessing the wind dates back thousands of years in human history. In many countries windmills have through the centuries been used for milling or pumping water.

In 1888, the American inventor Charles F. Brush built an automated wind machine connected to an electricity generator. This first wind generator could produce about 12kW of direct current electricity. Its rotor had a 17 meter wingspan and was made of cedar wood. The wind generator industry went through a great revival in the 1970s in the wake of the oil crisis.

Countries like Denmark, Germany, the Netherlands, Spain and USA have developed important wind generator industries and the Danes are able to provide about 21 per cent of their electricity from wind. Wind energy follows a third-power-law, meaning that increasing the wind speed by a factor of two will result in eight-fold increase in power. Wind maps have been constructed for most of the world and off-shore areas are a present target because they combine the benefits from high winds and reduced visual pollution.

Wind is on a winning path in Europe these days. I checked the information available on new power plants being erected in Europe in January 2005. The largest activity is in *gas* plants or about three quarters of all plants. *Wind*, believe it or not, comes in second place with 18,800 MW installed capacity being built which amounts to 13 per cent of all new plants erected (note that installed capacity in wind generators is always much higher than any average power production),

Unlike Brush's machine more than a century ago, the modern wind turbines are huge, or up to 5-6 MW in power including sleek rotors of up to 100 metres in wingspan, made of carbon fibre reinforced plastics and towering in the landscape. These beautiful structures are expected to be able to produce electricity at the sleek price of 0.02-0.04 USD per kWh at an installation cost as low as 0.7 USD per installed Watt.

In 2007 the installed wind harnessing globally was about 59 GW. Germany tops the list with about 21GW, Spain and USA follow and India seems to be surpassing Denmark as regards the number of new installations. The total installed wind generation facilities are now producing around one percent of the world's electricity.

In a publication by the Global Wind Energy Council, called Wind Force 12, there is an estimate for up to 12 per cent of the total global electric power stemming from wind in the year 2020.

Hydroelectric power

The Sun affects vaporisation of water on Earth and the subsequent precipitation. When this is combined with elevated land topography it holds the potential of hydroelectric energy. Water in an elevated mountain lake holds potential energy which can be returned and converted to kinetic energy as it falls down towards sea level.

An important electricity production stems from hydroplants where falling water is made to hit a rotating wheel, connected to a generator for the making of electricity. The design of Benoit Fourneyron of a water turbine in France in 1834 can be said to have led to the introduction of modern hydropower. By the end of the 19th century, water turbines had completely replaced the older waterwheels in most countries. Then, as the electric grids were developed, hydropower units got bigger and large hydropower installations have been built around the world to provide in many places an appreciable portion of the electric energy needs. The hydropower movement lost some momentum during the 1980s and early 1990s due to slackening of fuel prices. Political moves towards liberalisation of the energy markets have rekindled the interest for hydropower worldwide.

Countries like my own create up to 80 percent of the electricity used by this method. Figure 11 shows the most recent hydroelectric project in Iceland, the Karahnjukar project which is expected to yield 690 MW and has been Europe´s largest hydropower project of the new millennium. The project has caused enormous public debate in Iceland. Proponents advocate fossil-free energy for aluminium production and creation of employment in the region; whereas the opponents claim that the environmental footprint is simply not worth the sacrifice.

One of the advantages of hydropower is that the amount of CO_2 is much smaller than in the case of hydrocarbons. In fact, in order to prove any CO_2 emission we have to assess the life-cycle of a hydroelectric power plant and include in our calculation the concrete used in the construction of the plant, its reservoir walls etc. Anyhow, the hydroelectric power only leads to a fraction of the greenhouse gas emissions associated with fossil-based energy. The increasing problem of hydropower is associated with the reorganisation of land use and the creation of large dams.

Hydroelectric energy and other grid-connected renewable energy sources amount to about 8 per cent of the total energy used worldwide. This share is expected (International Energy Outlook 2006) to grow by about 1.9 per cent per year throughout the period up to 2025. Most of the projected growth in renewable electricity production is expected to result from large hydroelectric facilities in emerging economies. In Asia, for example, the need to expand the electricity production for a growing population with increased spending power outweighs environmental concerns linked to dams and reservoirs with the associated relocation of population centres. This is happening in China and Laos. In China, over a million people were moved out of the region affected by the Three Gorges dam project.

On the other hand, in transitional economies, additions to the hydroelectric capacity are expected to come from expansion or upgrading of existing power plants. In the most mature of markets hydroelectric resources have either been exploited or are situated quite far from population centres.

Today, hydropower provides around 20 per cent of the world´s electricity supply with the largest technically exploitable potential in Asia. Hydroelectric energy is about 7.1 per cent of total energy generation world wide. The energy hungry world needs different energy sources and the Energy Information Administration in its annual energy outlook, projects 5.1 per cent proportion of total energy in 2025 stemming from

Figure 11. The Karahnjukar dam in Eastern Iceland. Courtesy: National Power Company.

hydroelectricity. This represents a falling proportion. Other non-hydroelectric renewables are expected to increase their proportion, or from 2.2 per cent of electricity generation at the turn of the millennium, to some 3.2 per cent in 2025.

Biomass and Biofuels

Biomass is an important renewable on Earth. Usually, it is defined as plant or vegetation or agricultural waste useable for fuel or as an energy source. Households all the world over have through the ages used wood, peat or dried manure for burning to provide heat and light. In the previous section we linked all the renewables to the Sun. Biopower can be said to involve "solid sunshine", the harvest of the steady photosynthesis taking place in plants and algae. Biomass is the oldest form of renewable energy exploited by humans. When we consider biomass as a renewable it is assumed that "a tree is planted instead of a fallen one", unlike the case of coal and oil where there is no way of renewing the source.

Biomass energy conversion has both negative and positive environmental impacts: the burning of organic and fossil materials can emit harmful gases - while the disposal of agricultural and other organic waste utilises otherwise useless materials for obtaining energy.

The most common biofuel is ethanol. Pure ethanol has the chemical formula C_2H_5OH. Wheat or corn kernels are ground in a hammermill to expose the starch. The ground grain is mixed with water, cooked briefly and enzymes are added to convert the starch to sugar using a chemical reaction called hydrolysis. Finally, yeast is added to ferment the sugars to ethanol. After the process ethanol is distilled from the water mixture.

Brazil is the world leader in ethanol fuel with a production in 2006 about 16 billion litres and a half of the world production. All fuelling stations in Brazil sell pure ethanol (E95) and gasohol, a blend of 25 per cent ethanol and 75 per cent petrol. The U.S. is the second largest user with almost the same amount. The existing capacity in the world for ethanol production at the beginning of 2006 was about 34 billion litres (0.2 billion barrels) per year.

With crude oil prices reaching historically high levels, the global prospects for ethanol fuel use are growing. Ethanol production, derived from starch and sugar crops, such as sugar cane and cereals, is by some expected to expand tenfold in the period from 2005 to 2010. The direct pressure of oil prices on bio-derived ethanol (as well as food prices), as well as water usage, are causing some concern. In a recent article in *Scientific American*, it is argued that the production of one Megajoule of energy by corn ethanol requires 0.77 MJ of fossil energy. For cellulose ethanol as, for example, from corn stalks the required fossil energy would only be 0.10 MJ. On the other hand, the yield must be increased dramatically by the use of improved enzymes for the case of the stalks.

Sometimes the biomass can be used directly as a liquid diesel. The biodiesel is produced by a process called trans esterification of a variety of fatty acids, from vegetable oils; as well as fat by-products from the meat processing industry.

Engine producers have agreed on a standard called EN14214 for biodiesel for use in combustion engines. In 2003 the European Union issued a directive calling for biofuels to constitute 2 percent of all transport fuels by 2005, rising to 5.75 percent in 2010. The existing capacity in the beginning of 2006 of biodiesel worldwide exceeded about 2.2 billion litres per year. The biofuels today account for about 3 percent of the amount of petrol produced annually – but this figure is increasing.

There are a number of exciting new developments in the field of biofuels. In the marshlands of Orissa in eastern India and in the drylands of Gujarat in western India there is an ongoing experiment with planting the tropical Jatropha Curcas tree which bears oily seeds that can be used to make biodiesel. The plant is very sturdy and the seeds not edible; it can grow on poor land and reaches maturity in three years. The seeds have in the past only had limited value in folk medicine. The Indian project is expected to be able to utilise wastelands that have not been useful for practical food production and is expected to give millions of people useful work. The interest for the plant has been promoted by a British company, D1 Oils and, for example, Daimler.

As we go from biofuels to general biopower in more stationary applications, there are many ways of its utilisation. Through the use of *combustion* biomass for space heating, cooking or industrial processes. By heating water and creating steam the conversion to electricity is possible. A rule of thumb for electricity conversion is 1.6 kWh for every kilogramme of biomass. Small biomass power plants enjoy a lot of interest all over the world. *Co-firing* is another way of utilising biopower – here biomass substitutes for fossil fuels in an existing conventional power plant furnace. Compared to coal, biomass usually has less sulphur dioxide, SO_2, and nitrogen oxides, NO_x. The co-firing makes it possible to utilise biomass with the highest efficiency.

Moving to more advanced methods, *Pyrolysis*, is the process of decomposition of biomass at elevated temperatures up to 700°C in the absence of oxygen. The products from pyrolysis can be solids in the form of char or charcoal, or liquid in the form of pyrolysis oils. *Gasification* is a form of pyrolysis made with more air and higher temperatures in order to optimise the gas production. The gas can, in turn, be used to power internal combustion engines or turbines connected to an electric generator. *Anaerobic digestion* is a biological process by which organic wastes are converted to biogas – a mixture of methane (40-75 per cent) and carbon dioxide.

The cost of bioenergy varies a lot. It can be very low, from two USD cents per kWh for co-firing up to 10-15 cents in gasification plants. In countries rich in biomass like Finland, up to a quarter of the primary energy share is from biomass and nearly a fifth of the primary energy share for electricity production. The present capacity of biopower worldwide is close to 40 GW and is expected to grow by about 50 per cent up to 2010. The EIA assumes that biomass, including combined heat and power and co-firing in coal-fired plants, is the second largest source of renewable energy generation after hydroelectricity. The proportion of biomass in the world electricity production is expected to rise from a single per cent currently, to 1.4 per cent in 2025.

Geothermal energy

We will continue our journey through the renewables with geothermal energy. This fundamental energy has two major origins. Firstly, it stems from the enormous heat when the planets were formed by gravitational accretion and hot molten materials were left inside. Secondly, it originates from the radioactive decay of the various isotopes that were involved in the primordial interior of Earth.

The history of geothermal exploitation is not very old. Roman geothermal baths were well known in antiquity. In medieval times, deposits like sulphur where mined in geothermal areas like in Iceland. In the dark ages, one of Iceland´s main export items was sulphur for the making of gunpowder abroad. In the nineteenth century, a chemical industry was set up in the zone presently known as Larderello in Italy. This implicated extracting boric acid from geothermal fluids.

The boric acid was obtained by evaporating the hot fluids in iron boilers using wood as fuel. It was here that the Italian Francesco Larderel around 1827 had the ingenious idea of utilising the heat content of the geothermal fluid, instead of burning wood for the evaporation process.

At first the geothermal steam was used for raising liquids in primitive gas lifts and later in centrifugal pumps. In 1904 the first successful electricity generation from geothermal steam was realised in Larderello and in 1913 a 250kW electric plant had been developed. By 1942 the geothermal steam electricity production capacity in Italy had reached 127 MW. From then on New Zealand, California, Iceland and others joined the geothermal harnessing.

Most geothermal steam electricity plants are situated close to the source in hot geothermal regions. In a few cases the hot steam or the effluent from a combined heat and power plant CHP, is transported over long distances like in Iceland where a part of Reykjavik is heated by water led in a 30 km long pipeline from the Nesjavellir plant east of the city. Due to good heat insulation, the temperature drops only a degree or so on this journey. Iceland provides almost all space heating by geothermal energy and in

Reykjavik the Reykjavik Energy Corporation has displaced about 800,000 tons of coal which would have been needed to heat this northern capital city on an annual basis.

The geothermal energy can vary greatly in amounts, temperatures and forms. In the case of shallow magma intrusions, the temperature can be very high with the operating steam around 260°C. In these cases, flash turbines transfer the power to the electric generator unit. In Iceland, which is situated on the tectonic plate boundaries between the Eurasian and the American continental plates, magma is often at very shallow depths in the crust. In fact, there have been instances like in the Krafla region in northern Iceland where volcanic eruptions disrupted drilling for geothermal steam!

In other more common geothermal regions the temperature is lower. The operation of steam turbines obeys Carnot´s law and the efficiency of electricity production can vary from 14 per cent down to much lower values. The lowest values are obtained in low temperature areas where a second fluid is used by heat exchanger to power a separate system, commonly called a binary-cycle system. An Israeli company, Ormat Inc., is a pioneer in this field.

Also an ammonia based binary cycle named after its inventor, Alexander Kalina, and called Kalina, is used in the binary-cycle plants as for example in Husavik, Iceland. The primary water fluid temperature is considerably below boiling point. These systems are closed with no emission of the working fluid to the atmosphere.

The most economically favourable geothermal harnessing takes place in combined power and heat plants. With a combined power and heat plant, such as for example can be found in Iceland, the overall efficiency can be very high, even beyond 90 per cent when space heating of houses and dwellings utilise the heat left by the geothermal turbines. This is for example the case in my house in Reykjavik. I obtain geothermal water from the Reykjavik Energy utility, at 85°C. First the water is used for heating the house through water based circulation in radiators. Then the spent water is used for floor heating through a pipeline system and finally the 15-20°C hot water is led into a snow-melting system under the pavement outside the house. I have also experimented with the use of thermoelectric generation to add a few hundred watts to the electric energy portfolio of the home.

The exciting new development in Iceland is the plan to drill to greater depths in the hot geothermal zones close to the plate boundaries. The temperature is expected to become much higher, exceeding the critical temperature of steam and opening up possibilities for all sorts of chemical applications. For a given high temperature borehole it would be expected that the power of a typical hole could be increased by a factor of 5-10 by deep drilling. The estimated geothermal potential of Iceland, around 30 TWh on an annual basis, could then be increased dramatically.

While dwelling on this subject, let us consider how perfect and decarbonised geothermal harnessing really is. CO_2 emissions from geothermal plants are not insignificant but much lower than in the case of hydrocarbon burning. The geothermal fluids contain natural proportions of CO_2 and often some H_2S, CH_4 and fractions of other substances. The net CO_2 emission through a life cycle of a geothermal power plant will vary according to individual conditions. The World Bank assumes that geothermal en-

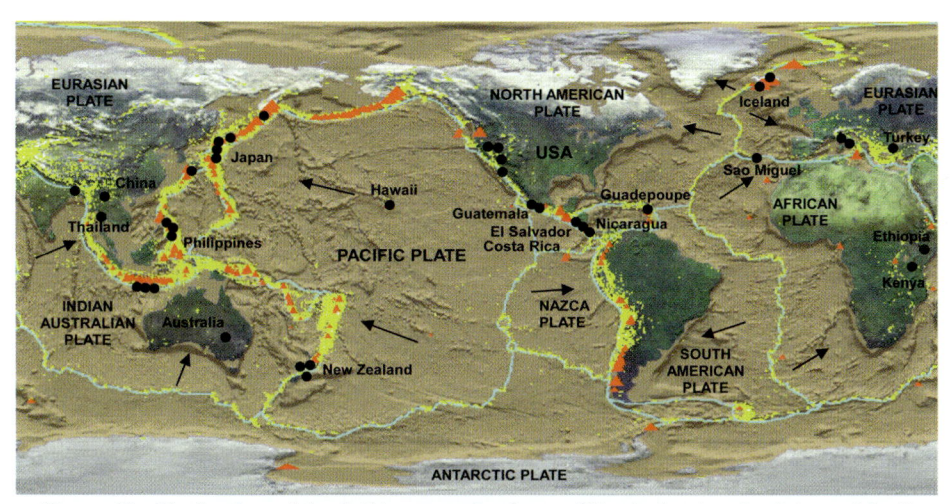

Figure 12. The borders of the drifting tectonic plates of Earth showing the "rings of fire", regions of high geothermal activity.

ergy world wide produces only about 5 per cent of the CO_2 which would be associated with a fossil fuel power plants of the same size.

The total installed geothermal power in the world passed the 8 GW limit around the turn of the millennium. The annual growth has been around 5 per cent. This type of energy has a net positive impact on the environment because of much lower pollution. Modern methods involve pumping the fluids back into the ground. The lifetime of geothermal plants has a theoretical upper limit defined by the heat content of the site.

A heat pump is a machine or device that moves or displaces heat from one location to another via work. Most often heat pump technology is applied to moving heat from a low temperature heat source to a heat sink of a higher temperature. Heat pumps enjoy an increasing popularity and will contribute to further use of geothermal power in the world. The total installed number of heat pumps in the world exceeds a million units and the total thermal energy involved is around 20,000 GWh. Sweden has been remarkably active in this area of geothermal utilisation.

Capital costs of geothermal utilisation can vary from around 3 USD per Watt to much higher. When hot, dry rock is harnessed with the associated pumping down of fluid to extract the heat, the cost can be appreciably higher. The cost per kWh can be as low as 5 USD cents in very large plants but is usually much higher.

Globally, geothermal energy is enormous. The heat content in the uppermost 5 kilometres of the Earth's crust represents more than 10,000 years of energy production currently taking place worldwide. This enormous energy amount can, of course, never be fully exploited.

Some regions of the world are very suitable for geothermal exploitation. Among these are areas covering about 0.2 percent of Europe: The Azores, The Canaries, Iceland, and the pre-Appennine belt of Tuscany and Latium in Italy, Aeolian islands, Aegean islands and Western Anatolia. The areas of Europe like Massif Central, Rhine Graben, Campidano Graben, Pannonian Basin and the island of Lesbos are of importance. Other parts of the world also count considerably. Figure 12 shows the borders of the continental drift areas, sometimes denoted rings of fires. Along the rings one can recognise various geothermal regions of the planet.

The U.S. Geothermal Energy Association predicts that in the next 30 years the potential capacity of geothermal energy will be 22 GW in South America, 10 GW in East Africa, 16 GW in Indonesia, 8 GW in the Philippines and 23 GW in the U.S. A total of 85 GW power generation is estimated over the next 30 years, about ten times the current installed capacity.

Wave and tidal power

Our journey around the globe can not be completed without mentioning three more ways of harnessing power which originate from wind, ocean currents and the movement around the Sun and of the Moon circulating the Earth and their pulling to and fro of the oceanic mass. My guess is that this form of power will receive increasing attention in the 21st century. To imagine the energy involved in tidal power one only has to visualise the enormous mass of ocean that follows the pulling of the Moon as it moves around Earth.

Seawater is some 800 times denser than air. The available energy for a given cross section of a water stream is thus 800 times greater than for air. This could make a basis for comparing a wind generator to a similar water based generation.

When estimating the global potential for tidal energy one must keep in mind that plants will be worthwhile only in places where the tidal range is particularly large, which means that our estimates do not involve a general assessment but careful site by site investigation. Using this approach, the World Energy Council has estimated the potential output of the most promising sites, considered to be about 386 TWh per year or about a tenth of that of hydroelectricity.

A world leading tidal power station built in 1966 is located on the Rance estuary, near St. Malo in Brittany in France. It has a peak production power of 240 MW. The only station in the US is the Annapolis Royal station (20 MW) in the Bay of Fundy established in 1984. Studies of available tidal power have been performed in many countries and have been published by the World Energy Council. In Argentina for example the estimated tidal power at the southern coast in a number of sites is over 37 TWh annually. In the UK, the Severn Estuary alone is expected to be able to yield up to 9 GW. In Russia, the Mezen Bay in the White Sea could yield some 15 GW.

Temperature gradients and salinity gradients in the oceans drive important global conveyor belts involving ocean currents. Ocean Thermal Energy Conversion (OTEC) is a means of converting into useful energy the temperature difference between surface water of the oceans in tropical and sub-tropical areas, and water at a depth of approximately 1,000 metres which comes from the polar regions. For OTEC a temperature difference of 20ºC is adequate, which embraces very large ocean areas, and favours islands and many developing countries. The continuing increase in demand from this sector of the world (as indicated by World Energy Council figures) provides a major potential market. I inspected a 210 kW station run by Louis Vega and colleagues in Hawaii 1993-1998, a very impressive demonstration project.

Wavepower involves an amount of energy which can be calculated as a fraction of the total energy dissipated by the ocean masses. The first wave-power patent was granted for a 1799 proposal by a Parisian named Monsieur Girard and his son to use direct mechanical action to drive pumps, saws, mills, or other heavy machinery. The world's first commercial wave energy plant, .5 MW, developed by WaveGen and is located at the Isle of Islay, Scotland. A company, Ocean Power Technologies, is building the first 1.25 MW commercial wave power station with their "PowerBuoy" system installed in the Bay of Biscay.

Installations of wave- and tidalpower have been built or are under construction in a number of countries, including Scotland, Portugal, Norway, the U.S., China, Japan, Australia and India. This technology is still in its infancy but has a lot of prospects.

Exotic future methods

We could carry on and mention more possible natural power sources but will leave it to a more specialised later book. We could, for example, mention the enormous losses of energy due to waste heat. Associated with the production of electricity on Earth there is, dependent upon methods used, typically twice more energy involved in the waste heat emitted. There are of course possible utilisation routes for waste heat, such as thermoelectric generators, but they seem too expensive with present technology. Thermoelectricity is about converting a current of heat into electric energy. This can be done by placing a suitable two-material couple across a large temperature gradient to form a junction. The physical effect is that the heat current borne by vibrations of the crystal lattice is converted into an electric potential drop across the junction.

The technology has only found applications in the space industry and specialised niches. In Iceland, for example, a thermoelectric generator powers a data transmitting station on Vatnajokull ice cap. It is powered by geothermal heat from a source in the glacier with the glacial surface acting as the cold side. A decade ago, I did some experiments with my colleagues to use thermoelectric electricity to produce hydrogen. That story will be told later.

We could mention the power of the plate movements and seismic energy but this will be tricky to harness properly.

Then, finally, we could mention the energy involved in lightning and the electric properties of the atmosphere. Lightning are in fact a derivative of solar energy: evaporation causes electric charge to be separated in the atmosphere with resulting huge voltage differences both vertically and horizontally between clouds and from clouds to Earth.

The instantaneous power of lightning around the globe outweighs the harnessed power of humans. Benjamin Franklin was reminded of this power when he got struck by a lightning through his kite line in the 1740s. Franklin must have been well insulated because the experiment can be fatal as was dramatically revealed when Prof. Georg Wilhelm Richmann of St. Petersburg Russia was executed when trying similar experiments in 1753. The State Senate House in Pennsylvania was equipped with the first lighting rod in 1752.

A typical energy involved in a single lightning is about 500 Mega Joule. If it takes place over a second, the power is half a GigaWatt. Some structures, like tall TV towers are struck up to a hundred times every year. There is, however, no simple way of harnessing this power. It awaits the future.

Our analysis so far seems to leave us with the fact that the traditional carbon related energies will continue to be the most important ones for the decades ahead of us. Harnessing of various renewable sources will continue to grow but is unlikely to fill the gap created by increased energy need. It is therefore of great interest to analyse what can be done to reduce the net carbon emissions and discuss decarbonisation, an action demanded in a transitional carbon economy.

Decarbonisation and cleaner energy

Having recognised the importance of hydrocarbons in the energy mix, and their dominance over renewables so far, we can make another observation. We note that hydrocarbons contain different ratios of carbon atoms to hydrogen atoms. Can something be done in order to reduce the carbon content? In the natural form, these sources contain a certain amount of carbon atoms for every hydrogen atom. To better understand the great importance of this ratio one can imagine that it also reflects the ratio of carbon dioxide to water in the emission of a burned fuel. The lower the ratio, the more water and less carbon dioxide in the emission. (See figure 13).

Within the ancient to present societies which were and are burning wood, the ratio of carbon to hydrogen in the fuel is close to about ten carbon atoms to every hydrogen atom which is characteristic of the wood composition when water has been removed (10:1).

In simple coke this ratio is about 6:1. In coal the ratio is slightly in favour of carbon over hydrogen, making the ratio 2:1 in total.

This ratio is even more favourable to hydrogen in conventional oil. In petrol and diesel, containing the chains of hydrocarbons as we discussed earlier, the ratio is close to 1:2. Here, the ratio is in favour of hydrogen.

Natural gas has a simple ratio of 1:4; four hydrogen atoms to every carbon atoms, as the formula is CH_4.

If we arrange the carbon to hydrogen ratio in the history of fuels as a function of time (figure 13), we can see a trend of ever lowering ratio. Only in the case of pure hydrogen, does this ratio converge to zero and the only emission is water. From these considerations it seems that opting for natural gas instead of coal would in general increase the decarbonisation and decrease emissions.

Nature has chosen to use hydrogen in its binding of solar energy in the form of hydrocarbon production. Nature gets rid of, or stores, the hydrocarbons by deposition deep in Earth. With the advent of the human race, these hydrocarbons were taken to the surface and returned to the atmosphere in a very non-sustainable manner in a sometimes untamed burning. Humans seem to have evolved always burning something!

Equipped with the understanding of the energy conversion system and the nature of CO_2 emissions we could attempt to suggest ways to *reduce* them. This is an immense

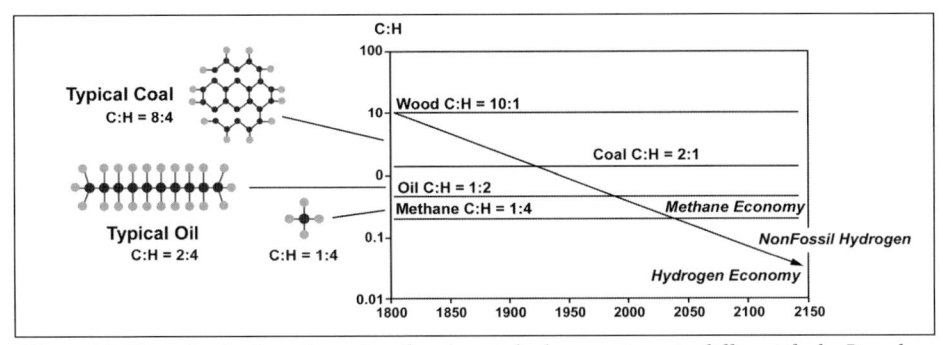

Figure 13. Decarbonisation, the ratio of carbon to hydrogen atoms in different fuels. Based on Ausubel.

task because it will have to touch upon all aspects of energy use and lifecycles of energy plants and infrastructure. Let us deepen this for a moment.

As regards available imminent actions, *increased energy efficiency* seems to be the most natural first step towards reducing the CO_2 associated with a given use of hydrocarbons, such as driving a hundred kilometres or producing a kilowatt-hour of electricity. First we note that the present level of carbon pumped into the atmosphere is about 7 billion tons annually. If nothing is done this level will rise to about 14 billion tons (or Gigatons) in 2055. The concentration of CO_2 in the atmosphere is currently about 375 ppm, parts per million. At the present rate of emissions CO_2 is increasing by about 1.5 per cent annually. The figures used here assume that 60 per cent of the carbon dioxide is absorbed in the atmosphere and 40 per cent taken up by the oceans and the biosphere. Many environmentalists have recommended that the aim should be stabilisation below doubling or at around 450-500 ppm. So where could the efficiency improvement start?

An obvious choice seems in the transport sector. In 2055 there will probably be 2 billion cars on the roads, three times more than at present. If the trend set by Toyota and others with their hybrid cars continues, it is conceivable that an average car would run twice as long a distance on a full tank compared with today. Pakala and Sokolow at Princeton University, New Jersey, have pointed out that by this efficiency increase, a step towards reducing the CO_2 emissions would certainly be taken with a resulting projected one Gigaton of reduction in emissions (on a 50 year basis), a step towards meeting the enormous challenge of the horrific 14 Gigatons mentioned earlier. In their paper, they define a "Wedge" as a strategy to reduce carbon emissions that grows in 50 years from zero to 1.0 Gigaton of carbon per year. Figure 14 shows how the wedges approach would slow down the growth of the emission curve.

The first wedge they recommend is energy efficiency of transport. They also suggest that a sociological change would be needed in order to *reduce the reliance on motorcars and increase efficiency of energy use in buildings*. Thereby they obtain additional two Gigatons of avoided increase before 2055. The latter is through heating, cooling and refrigeration improvements.

An additional Gigaton would be saved by *converting coal powerplants*, presently running at about 40 percent efficiency, to 60 percent. A quarter of the present emis-

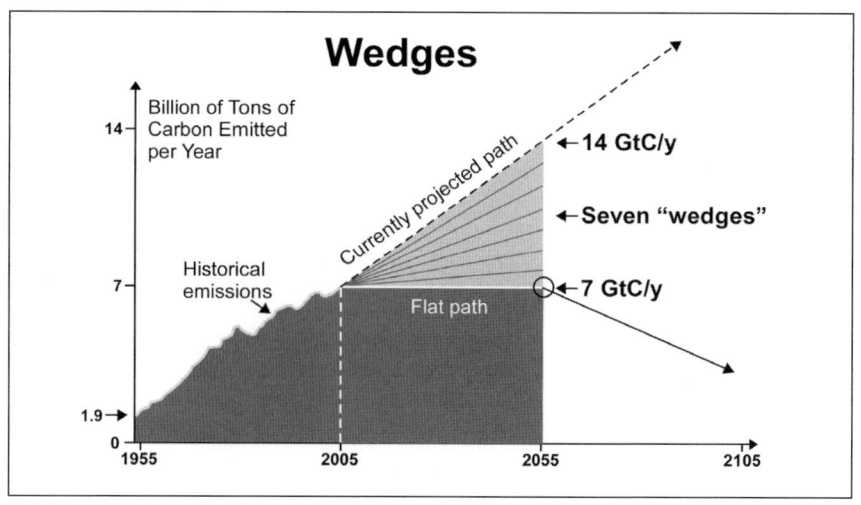

Figure 14. Wedges. How to wedge down the carbon contribution from various energy sources. Based on Pakala and Sokolow.

sions into the atmosphere stem from those power plants which run on 32 per cent efficiency.

A further Gigaton reduction could be obtained by *decarbonisation of power plants and a conversion from coal used presently to natural gas.* Pakala and Sokolow then go into suggesting the so called precombustion capture of CO_2 in power plants and a series of carbon sequestration actions. One of these would be to make Fischer-Tropsch fuel out of emissions and hydrogen and thereby make the CO_2 rerun through the energy cycle. They also point out *that maintaining the same rate of introducing new nuclear plants* as in the last decades of the 20th century, yet another Gigaton of reduction would be achieved.

So what is their anticipated role of renewables? The first renewable resource they suggest is *wind electricity*. Their calculation in order to obtain one Gigaton reduction results in the need for increasing wind generator capacity on Earth by a factor of 50 by installing two million wind generators of the one Megawatt size until 2055. This shows us the enormity of the problem and the challenge it possesses and how difficult it will be to replace the old fossil energy base with renewables.

The second renewable source addressed by the two scientists is photovoltaics. According to their calculation to obtain a Gigaton reduction, there will be a need to *increase the present 3 GW photovoltaic production capacity* by, on average, two to three square metres of photovoltaic cells per capita for the whole human race!

Wind generators coupled with hydrogen production by electrolysis of water is yet another suggestion by the Princeton team. *The main displacement would be petrol and diesel by hydrogen.*

A further Gigaton of reduction would be obtained by increasing the *biofuel generation* to about 34 billion barrels daily, an increase to about 50 times the present production rate. This is quite a challenge for most countries with large bioproduction.

In the elegant paper by the Princeton team it is pointed out that by *reducing deforestation*, more than a Gigaton of CO_2 reductions can be achieved. The same is suggested as regards *agricultural soil management*. I spoke to Robert Sokolow in New York some time ago and he noted that an additional *methane management package involving* landfill gas, manure etc. would also go a long way towards defining a wedge.

Carbon dioxide Capture and Storage (CCS), is a process consisting of separation of CO_2 from the industrial and energy related sources. This is followed by transportation to a storage location and long term isolation from the atmosphere. This can be done by various methods most of which are in their infancy. The most obvious capture of CO_2 can be applied to point sources which, following a compression, can be transported for storage in geological formations, in the ocean, in mineral carbonates and/or for use in industrial processes. The most common methods involve post-combustion, pre-combustion and oxyfuel combustion. The method mentioned last is in a demonstration phase. Pre-combustion involves reacting the fuel with oxygen or steam in a catalytic reactor to give CO_2 and more hydrogen. The CO_2 is then separated and the hydrogen used as fuel in a gas turbine combined cycle plant.

An important point to make here concerns sequestration steps that the oil industry can take closer to the origins, usually termed *upstream*. There are emerging examples of such steps. The oil industry can already influence the CO_2 balance to some extent by action upstream – close to the sources. In a report from 2005, Bellona, a non-governmental-organization of Norway has recommended that the Norwegian government establish a new and original value chain for CO_2. Two corporations – one for capture and the other for distribution and sales of 17 million tons of CO_2 annually - are suggested to be established. The operation is believed to need an investment of 86 billion Norwegian kroner which is roughly 15 billion USD.

The methodology of the Bellona task group is the following: In oil production the natural gas pressure of an oil field is used to lift the oil to the surface of the field. The Norwegian idea assumes that about a third of the resources will be tapped. As the source is reduced, pressure falls. In the next phase gas and water would be pumped down to maintain pressure. In this way it would be expected that up to half of the reservoir would be tapped.

In a proposed third phase, extra CO_2 is injected, and this would wash out the oil which otherwise would be left in the field for ever. It is estimated by the Norwegian group Bellona, that 5 percent more oil would be recovered in this third phase, resulting in a value of 1,000 billion NOK (175 billion USD) in this final phase using today's value of oil on the world market on a 40 year basis. The philosophy is usually named "enhanced oil recovery".

The Norwegian idea is to build five gas cleaning utilities which would be deposited in two regions connected through a pipework. One in the Norwegian Sea and Draugen field deposition areas offshore and one in the North Sea with the Heimdal field as a main deposition area. This is a great project which is expected to sequester some 16 million tons annually and could be an example for the world to follow. British Petroleum is planning a natural gas power system associated with the Miller field in the North Sea with Carbon Capture and Sequestration CCS amounting to 1.3 Megaton CO_2 annually.

The pumping back into oil fields is the simplest model suggested so far. In a more sophisticated model suggested by Julio Friedmann at the Lawrence Livermore Laboratories in the US, a hypothetical 1000 MW coal power plant injects 60 Megatons of CO_2 over 10 years into underlying geological formations of 10 percent porosity. The horizontal area of the geological formation is 40 km² at a depth around 2 kilometres.

A much more risky method would involve ocean storage where CO_2 is injected and dissolved in a water column below a kilometre of ocean depth. Below 3 km CO_2 is denser than water and would be expected to form a "lake" that would delay dissolution of CO_2 into the surrounding environment.

On the international scene it will be interesting to follow the development of the Carbon Sequestration Leadership Forum (CSLF) initiated a few years ago – this forum of 21 countries is an excellent vehicle to influence decarbonisation world-wide. Governments of the world are realising that carbon sequestration is a dead serious issue. The Carbon Sequestration Leadership Forum is an international climate change related initiative that is focused on development of improved cost-effective technologies for the separation and capture of carbon dioxide for its transport and long-term safe storage. The purpose of the CSLF is to make the existing and expected sequestration technologies broadly available internationally; and to identify and address wider issues relating to carbon-capture and storage. This could include promoting the appropriate technical, political, and regulatory environments for the development of such technology.

In 2004 The Earth Institute of Columbia University in New York initiated The Global Roundtable on Climate Change GROCC. Professors Jeffrey Sachs, Klaus Lackner and Wallace Broecker of Columbia are among the leaders of the GROCC. Klaus Lackner has proposed an ambitious idea where ambient air is collected around the world as a first step to capture CO_2 for further sequestration. The roundtable contains up to one hundred international and U.S. corporations. The aim of the GROCC involves socioeconomic analysis, technical solutions and future foresight. A technical committee under GROCC is examining ways feasible for capturing and/or sequestering CO_2.

As a result of this work, experiments are taking place in Iceland where CO_2 is pumped into basalt in a project led by S. Gislason.

Fig. 15. Klaus Lackner's ambient CO_2 collectors rise as huge trees in the landscape. Source: Stonehaven Productions; Lackner.

Sharpening the pencil

In the previous pages we have devoted a whole chapter to describing the primary energy sources available. We have also addressed the main problem associated with the use of fossil fuels and the capacity of non fossil resources to replace them. It seems that humankind will have to be prepared to live with the existing fossil fuels as a primary resource for the next half to one century

However, as we have pointed out, there are many ways of increasing energy efficiency and continue green house gas reduction. The prospects of *non-fossil* resources are very good as far as they reach, although not being able to fulfil the total energy needs. Therefore, we have to expect an interim period, a rather long period, with fossil energy as a main source. During this present century it can be expected that the role of non-fossil energy can increase significantly. Governments will continue to exercise tax incentives to curb the green house gas emissions. Carbon sequestration is a must.

What will then replace heat engines and the cumbersome conversion of hydrocarbons into electricity and motion? We have seen the problems of low energy conversion efficiency and the accompanying CO_2 emissions. The energy efficiency of the so called fuel cell will be studied in the next chapter. With the present knowledge, hydrogen can be bound and stored in a more sustainable way than hitherto. Additionally, with the advent of fuel cells, hydrogen can be harnessed in a way which makes the efficiency much higher than in the case of the Carnot era where hydrogen containing compounds were burned and their combustion energy utilized by some sort of effectively "moving a piston". By using the revolutionary *fuel cell* a step is taken into a post-Carnot energy era; from burning to utilisation of the free energy of the electrons. This is what I mean with the phrase *taming of the proton* in a way which was realised with the advent *of fuel cells*.

One important aspect of the future is linked to energy carriers. In an energy economy devoted to increased efficiency there is a definite potential candidate to assist the transition from fossil fuels with the aim to gain ever increasing importance. This candidate is HYDROGEN. With fuel cells we can expect hydrogen to be a prime energy carrier for efficiency and the utilisation will involve the chemical energy of hydrogen rather than the heat of combustion as translated into the Carnot-natured combustion engines. Hydrogen combines with oxygen to form water. Renewability is clear.

In the next chapter we intend to study the ways of utilising hydrogen as an energy carrier, the production of hydrogen, its compression and storage, and finally we intend to devote a whole section to the post-Carnot era of fuel cells.

Through all of the 20th Century Humankind has harnessed **electrons** and introduced them into technology with electric motors, electric dynamos and microelectronics. This transition took over a century and is still ongoing. Now, the turn has come to **protons**. *In a future hydrogen and fuel cell energy economy, the proton has been tamed and harnessed and will be the next energy revolution on Earth. It will be harnessed with fuel cells and put to do work in electric motors.* This is the main message of the present book and we will start working on the subject in the next part after a short introduction to the history of the proton.

PART III
THE TAMING OF THE PROTON

Discovering the proton

Once upon a time in Greece there was a great philosopher called Thales from the Ionian city of Miletus. The historian Herodotus mentions Thales in his historical notes and his great achievement in foretelling an eclipse, now generally agreed to have occurred on May 28th 585 B.C. According to another historian Diogenes Laertus, Thales died during the 58th Olympiad (548 B.C.) aged 78. (See figure 16).

Thales is often described among the "Seven Wise Men" of ancient Greek philosophy and is universally held to have introduced geometry into Greece. He gave advice to the Greek authorities regarding unification of cities, diversion of rivers; and even the general guidance to navigators to steer by the Little Bear star sign rather than the Great one. Thales was so practical that, according to Aristotle, he was perhaps the first "takeover businessman" when he hired all the available olive presses in his region and was, when the season came, able to charge his own price for reletting them.

Aristotle regards Thales as the founder of European philosophy. Moreover, Aristo-

Figure 16. Thales of Miletus.

tle tells that he was the first to suggest *a single universal substratum*, namely water. And water was supposed to be the basis of the manifold of forms of material objects.

For the Greek atomists Leocippus and Democritus, who lived about a century after Thales, his main ideas about the underlying unity of matter remained very important and have, as such, survived as a core concept to the present day.

Twenty four centuries later, in the eighteenth and nineteenth century, scientists started to narrow the search for the truth about the atomic nature of matter. In 1671 Robert Boyle described the chemical properties of hydrogen and five years later the element was first isolated by Henry Cavendish. A prominent French scientist and revolutionary, A. Lavoisier, was the one who created the name hydrogen around 1788. The guillotine in Paris ended his life before he could enjoy the benefits of his creation. The execution had nothing to do with the element. The idea with the name is to connect the Greek word for water "hydro" and "genes" which means "born of". In the German language the name of hydrogen is "Wasserstoff" which means exactly the "stuff in water". Five years before Lavoisier's name-giving, a French physicist, Alexandre Cesar Charles, launched the first hydrogen balloon flight and flew himself in a manned balloon the same year.

British chemists, Sir Humphry Davy and John Dalton led the way by important experiments and a young physician from Edinburgh University, William Prout (1785-1850) started to measure the weight of such atoms as iodine, phosphorus and sodium, iron, zinc, potassium and beryllium. Prout used Davy's result for the atomic weight of hydrogen and proposed that all matter was constructed with the hydrogen atom as a fundamental building block. Prout was so close to the right masses measured, that many scientists believed he had found the ultimate holy grail of chemistry.

As modern science knows, the uncertainty in Prout's calculation stems for example from the comparatively very light mass of electron and the dominance of the mass of the nucleus in the mass of the hydrogen atom. The electron, which was discovered by J.J.Thomson in 1897, is so light that it contains only one part in 2000 of the mass of the proton, the constituent of the hydrogen nucleus. Thomson believed that the electron was embedded in the atom like a raisin in a cake; but with the discovery by Ernest Rutherford in 1911 of the atomic nucleus and the subsequent work of his young assistant Niels Bohr, a new picture emerged. In this model, a negatively charged electron orbits about the positively charged proton nucleus. The name proton first appeared in print in 1920 and was coined by Rutherford (1871-1937; fig. 17) who was using the Greek *proton*, neut. of *protos*, "first". (Earlier the word had been used in embryology).

The full picture of protons and the uncharged neutrons occupying the nuclei of other atoms did not appear before J. Chadwick's discovery of the neutron around 1932.

After the First World War which to some extent delayed the progress of the atomic search, Thomson's colleague F.W. Aston built a mass spectrograph and was able to not only measure the mass of atoms but also show that the nucleus of a given element could have different masses and the concept of isotopes was founded. H.C. Urey discovered in 1931 heavy hydrogen, deuterium, the isotope containing a proton and a neutron in its nucleus. There are three isotopes of hydrogen, the normal hydrogen; deuterium with one neutron accompanying the proton in the nucleus; and finally tritium where two

Figure 17. Ernest Rutherford.

neutrons and a proton dwell in the nucleus. The normal hydrogen accounts for 99.85 per cent of all hydrogen atoms found in nature; Deuterium accounts for 0.015 per cent. Tritium is an unstable isotope (fig. 18).

Is there an underlying theory to explain the various types of hydrogen in the form of different nuclei? Yes, indeed. For completeness we mention lastly that in order to fulfil the laws of physics, the forces of the nucleus are provided for by the existence of even smaller sub-atomic units: quarks and gluons, the name of the latter so wonderfully related to "gluing together" nuclear particles. This is basically the light in which modern physics sees the atom. If we break the quarks and the gluons further up, we reach the stage where these resulting entities will be composed of superstrings, whiskers of the most fundamental building units of matter, which behave almost like feathers filling an infinitesimal pillow and bringing us back to the first moments of the world we discussed earlier.

So humankind had come a long way to the modern picture of hydrogen: from a hypothesis by Thales about water as a primordial substance, to a model of hydrogen

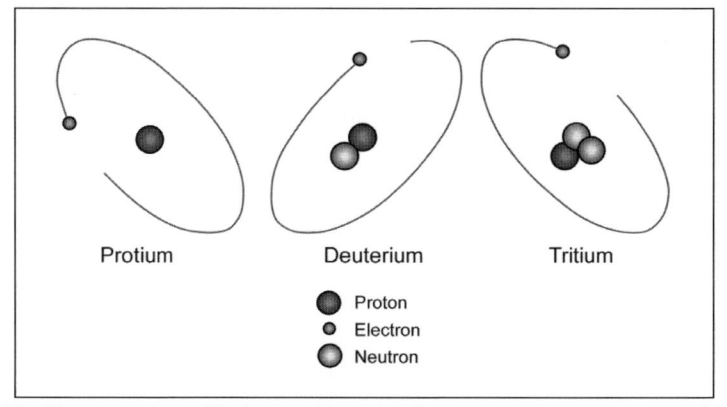

Figure 18. Three isotopes of hydrogen. Protium, the most common isotope involving about 99.85 per cent of all natural hydrogen atoms; Deuterium or heavy hydrogen which amounts to 0.015 per cent and finally Tritium, an unstable isotope.

and finally to the understanding of its building blocks: of protons, electrons and neutrons.

To fulfil our aim to harness or tame the proton, we will start by looking at some fundamental properties of hydrogen. Then we will go on to how it is produced and stored and finally culminate in the utilisation of hydrogen as an energy carrier.

Some key properties of hydrogen

Hydrogen, which from now on we will denote with H_2 indicating the fact that there are two hydrogen atoms in a molecule, is a colourless and odourless gas at atmospheric temperatures and pressures. It is very flammable but yet non-toxic and ignites in the presence of oxygen and burns in air with a pale blue, almost invisible flame. It is the lightest of all elements and approximately one-fifteenth as heavy as air. It ignites very easily and forms, together with oxygen or air, an explosive gas mixture. The speed of the chemical reaction can vary with geometry and amounts of materials and in extreme cases we would talk about deflagration rather than explosion.

In fact, hydrogen has the highest combustion energy release per unit of weight of any commonly occurring material. This property of course makes hydrogen a fuel of choice for demanding combustion tasks such as in space rockets.

On the other hand, if hydrogen is cooled down, it remains a gas until it reaches about -252.76 °C, or about 20 degrees Kelvin. The only element with a lower boiling point is helium which boils at only 4.2 K. In the liquid state hydrogen is again a transparent, odourless liquid that is no more than one-fourteenth as heavy as water. Liquid hydrogen is not corrosive or particularly reactive but when converted from liquid to gas it expands approximately 840 times.

Cyber-Appendix I shows the various physical properties of hydrogen. One unusual behaviour of hydrogen concerns its deviation from the so called ideal gas. For an ideal gas the relationship between pressure and density at constant temperature is linear. As the pressure is increased, the volume decreases in inverse proportion. In the case of hydrogen there is an inherent springy behaviour and, as the molecules are compressed together, their density does not follow linearly. This deviation from linearity is about 6 per cent at pressures around 100 MPa, equivalent to thousand atmospheres of pressure.

Figure 19. shows the so called phase diagram for hydrogen. This diagram shows as a function of temperature (vertical axis) and density (horizontal axis) the various phases of the material. In most cases of normal use as we know it, hydrogen is a gas composed of two atoms per molecule. Below its boiling point, hydrogen is indeed a very cold molecular liquid.

Going into extreme parts of the phase diagram, new things crop up. Looking in the uppermost, high temperature area, one can see the occurrence of the plasma phase; very much in line with what we mentioned in the chapter on the original state of the universe. As we lower the temperature we enter the gas phase and if we go even lower we enter the area where hydrogen is solid. Depending on density, hydrogen can be either a molecular solid or a metallic solid. In the metallic solid hydrogen has been compressed to a stage where each electron becomes a free conducting electron.

Planet Hydrogen 53

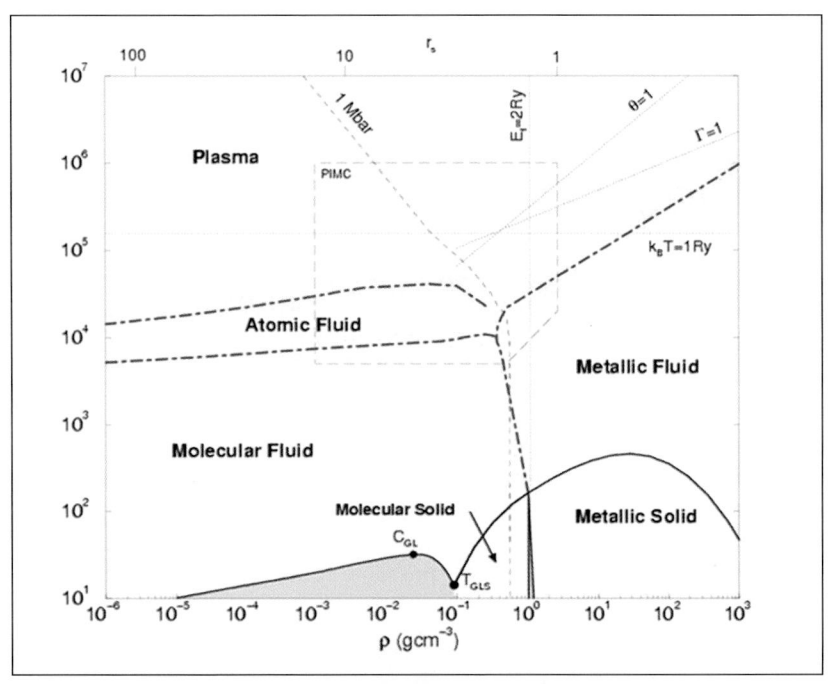

Figure 19. The phase diagram of hydrogen. The horizontal axis shows the density. The vertical axis denotes the temperature in Kelvin. Based on various sources.

The quest for *metallic* hydrogen is an old dream of physicists. Eugene Wigner advanced such a theory in 1935, but for a long time it was difficult to verify experimentally. There was no limit to speculations about its existence in heavenly bodies and for example in the interior of Jupiter there was the expectation to find hydrogen in the form of a superheated metal.

In a remarkable experiment by Weir, Mitchell and Nellis reported in 1996, a team from the Lawrence Livermore Laboratory used a 20 metre long two-stage light-gas gun, originally built by General Motors in the 1960s for ballistic missile studies, to create shock compression of hydrogen. The projectile runs at up to 8 km per second and compresses the hydrogen to enormous pressures and temperatures. The Livermore team found a transition from a semiconducting diatomic fluid to metallic phase at 3000K and a nine-fold compression of the initial fluid. Other teams have since the original experiment confirmed the metallisation by more classical approaches using a so called diamond anvil cell and Wigner's theory is seen as verified.

Coming back to Jupiter which is about 90 per cent hydrogen, it would be expected that its interior is made up of metallic hydrogen, which in fact also may be present in large planets outside the solar system. Now, since Jupiter has a magnetic field of its own, it is likely that the charge current causing the field spins around inside the planet.

The Livermore team has speculated about the possibility of their results becoming important to further the progress of hydrogen fusion experiments needed to make nuclear fusion a feasible technology.

PRODUCING HYDROGEN

A variety of routes are possible in order to derive hydrogen from a diverse range of primary feedstock, including fossil fuels with emphasis on natural gas, but also oil and coal; from biomass and from water. The primary energy resources also vary greatly. The main criteria for production are capital and maintenance costs, efficiency, flexibility in design and operation, safety and hazard risk management – all including the demand that waste minimisation is always kept in mind. To meet this last requirement a so called life-cycle-analysis is performed on every pathway studied. Life Cycle Analysis (LCA) is a technique for quantifying and assessing the inputs and outputs affecting environmental performance associated with a product throughout its life cycle from production, to use, and to disposal. Life Cycle Analysis can assist in identifying opportunities to improve environmental performance.

Hydrogen production routes can be divided into a number of categories: Fossil fuel and chemical; renewable energies and electrolytic processes; biological systems; and nuclear fission or fusion incorporating electrolytic production or chemical methods. We will now examine these routes further. Figure 20 shows the variety of pathways possible and Figure-table 21 shows the same concept from a point of view of raw materials.

Hydrogen from fossil-fuels and chemical processes

The elements hydrogen, carbon and oxygen are closely interrelated in the elemental mix of existing fossil deposits. When we discussed decarbonisation we also saw this relationship even further through the ratio of carbon to hydrogen. A one-to-four ratio exists in every molecule of methane, CH_4, the main constituent of natural gas, where there are four hydrogen atoms linked to a carbon atom.

The most economical way of producing hydrogen today is based on production

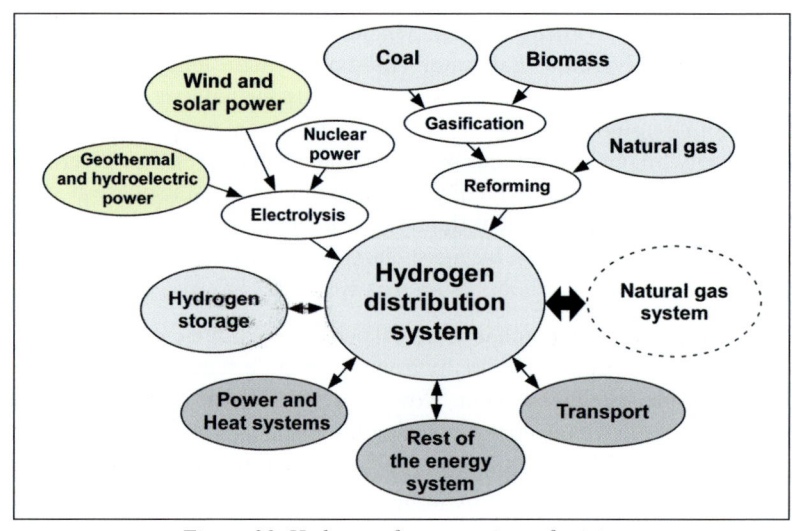

Figure 20. Hydrogen from a variety of origins.

Planet Hydrogen

		Primary feedstock	Other feedstock	Energy	Quantity on earth *Gigaton level*	Carbon footprint or emissions
THERMAL	Steam reforming	Natural Gas	Water	Heat	345 (USGS 2000)	Carbon sequestration needed
		Oil		"	400 (Smil 1987)	
	Gasification Pyrolysis Thermo-chemical	Coal	Water, Oxygen	Steam at high T and P	6200 (Rogner 1997)	Carbon sequestration
		Biomass		Moderately high temp.	Laherrére 2000)	Essentailly renewable
				Steam		
		Water		Nuclear	1.400 000 000 USCS incl. Oceans	Nuclear waste disposal problems
ELECTROCHEMICAL	Electro-lysis	Water		Renewables incl. Solar wind, hydrelectric	1.400 000 000 USCS incl. Oceans	Emissions mostly related to Life Cycle
	Photo-electro-chemical	Water		Direct sunlight	1.400 000 000 USCS incl. Oceans	Minor emissions
	Photo-biological	Water	Algaie strains	Direct sunlight	1.400 000 000 USCS incl. Oceans	No emissions
BIOLOGICAL	Anaerobic digestion Fermenta-tive micro-organisms	Biomass		High temperature steam		LCA related minor
		Biomass		High temperature steam		LCA related minor

Figure 21. Hydrogen production methods and the demand for raw materials (Sigfusson Royal Society Lecture).

methods from fossil resources such as the so called reforming of natural gas. The methodology is quite old. In the fruitful years of the middle of the nineteenth century, oil exploration and the production of *town gas* – a mixture of hydrogen and carbon monoxide- by the passing of steam over glowing coal, were developed almost simultaneously. The oldest such systems were established in Paris in 1837 and spread quickly through the world. In sub-arctic Reykjavik, by then a capital of a developing country, town gas was produced from 1910 to about 1956 and fed via pipeline to residential houses in order to power the cooking stoves, and to streets for lighting. The technology was gradually replaced with the advent of electric cookers and the use of electric bulbs.

The making of town gas was the first example of a large scale method for converting a fossil fuel into hydrogen. The various processes possible and chemical reactions needed for isolating the hydrogen component of hydrocarbons are shown in detail in Cyber-Appendix II. Basically they involve a reaction between hydrocarbon chains, C_nH_m and water H_2O where the chemical product becomes CO_2 and H_2. We will now spend some time looking at the important methods and implications of making hydrogen from fossil fuels.

The world production of hydrogen is about 50 million tons annually and most of this amount comes from reforming natural gas. Most of the hydrogen produced is used in refineries and is, as such, a crucial ingredient in the oil industry. The hydrogen is produced mostly by a process called *steam reforming*. The basics of this process involve heating the mixture of natural gas (methane) with steam; in turn they react to produce carbon monoxide and hydrogen. Then the water combines with CO to form CO_2 according to the so called "water-gas shift reaction".

The reactions are not complete as they are written, and some natural gas and carbon monoxide goes through without reacting. At this point, a catalyst is used for burning the product in the presence of oxygen from air. This burns most of the CO to CO_2. A sophisticated engineering is used to further purify other unwanted compounds from the original feedstock, such as sulfur. It is very important, as we will see later, to eliminate CO completely, since this is a contaminant when used in a fuel cell and will gradually reduce its lifetime.

A methanol (CH_3OH) reformer combines the compound with water vapour in presence of a catalyst which splits the methanol into CO and H_2. Again here, the water vapour reacts with the CO to form CO_2 and H_2. The reader is referred to the section on efficient hydrogen utilisation to discuss direct methanol fuel cells which enjoy special attention in hydrogen technology.

The technical hurdles in the steam reforming process of natural gas, or for that matter other hydrocarbon streams, involve getting rid of naturally occurring sulfur-containing compounds which stem from the ancient origins of the fossil fuels. The catalytic processes convert most of the sulphur containing compounds to hydrogen sulfide (H_2S), the infamous "rotten egg smell" compound. The H_2S is then absorbed in a zincoxide (ZnO) bed and converted to zinc sulfide (ZnS) releasing the hydrogen. Further purification is done by a copper based catalysis.

After the purification, steam is added to the feed in an expanding mixture which consists of various molecular components explained in Cyber-Appendix II, mostly hydrogen and carbon monoxide. Pressures can be high, of the order of 20-40 atmospheres and temperatures in the range up to 950°C. The products from the reformer include CO, CO_2 and CH_4. The so called water gas shift reaction which converts CO to CO_2 (and still involves more use of steam!) is used for eliminating carbon monoxide, and finally the important "chemical scrubbing" is employed in order to clean the mixture and isolate the hydrogen.

Yet another method is partial oxidation where pure oxygen is used in stead of steam in order to tackle the hydrocarbons, convert them into carbon monoxide and hydrogen. This process we describe in more technical detail in Cyber-Appendix II. It is mostly used for heavy feeds. A third method is the *auto-thermal reforming (ATR)* process which is a mixture of reforming and partial oxidation. Overall, the partial oxidation and ATR are the most compact of the reforming technologies.

Hydrogen from natural gas is the cheapest method available world wide. The US Department of Energy has set a target for 2010 for the cost of producing hydrogen to just over one USD per kilogramme H_2, at a primary energy efficiency of 75 per cent. This is and continues to be the lowest hydrogen cost targets considered (1.06 USD/kg).

Bear in mind that all these methods of converting fossil fuels into hydrogen suffer from one fundamental problem: they each and everyone emit CO_2 to the atmosphere and are, no different from the burning of fossil fuels in power plants, automobiles etc., all notorious greenhouse gas emitters as well as having problems with associated volatile compounds. The vicious cycle which is created in this way is not halted unless there is a way to trap the CO_2 early in the process, by sequestering the carbon dioxide as we have previously mentioned. A final price of hydrogen from fossil fuels must take into account the cost of sequestering CO_2.

Hydrogen from biomass

Biomass is a very heterogeneous and chemically complex renewable resource. The term biomass encompasses any plant, animal (or humans for that matter) derived organic matter. Biomass and microorganisms in the biosphere are important as regards carbon and hydrogen content. Similarly to the methods we discussed in the previous section, hydrogen can be produced from renewable energy sources like various forms of biomass. However, unlike hydrocarbons from fossil resources, plant biomass is termed renewable because the carbon dioxide released to the atmosphere by the biomass gasification was previously absorbed from the atmosphere and converted to hydrocarbons by photosynthesis of the plants involved.

The feasibility of producing hydrogen from biomass is very much dependent upon local conditions. Most regions of the world outside the polar regions have ample sources of biomass to serve as a feedstock for such production. They range from woods, to agricultural crops, agro-industrial waste, animal manure, sewage sludge and household waste.

The methods available for the hydrogen production from biomass are discussed in Cyber-Appendix II, but can perhaps be divided into two major routes: *thermochemical and biological processes.* We will start by considering the thermochemical processes.

When addressing thermochemical conversion, remember that when we discussed decarbonisation, the proportional amount of hydrogen in the biomass varies for the different feedstock. As an example, wood contains cellulose which has about 6 per cent hydrogen. When this promising hydrogen containing substance is converted, there arises the need for getting rid of volatile and dissociated components contained in the wood. These dominant ingredients (up to 4/5 of the mass) are vaporised at temperatures between 500-700°C. There exist sophisticated pyrolysis methods and technologies for use in gasification of wood-like biomass. The word pyrolysis comes from the Greek "pyro" for "fire" and "lysis" for "cutting". Many believe that such conversion, although expensive at present, will be a major technology for a future hydrogen economy, enjoying a lot of research interest. Pyrolysis of biomass involves heating the biomass up to very high temperatures in the absence of oxygen or water to form a so called bio-oil. This is subsequently converted to hydrogen via catalytic steam reforming and shift reaction.

In Vaexjö in Sweden, a number of partners including the local university, have been setting up a gasification testing facility with 18 MW thermal energy corresponding to some 4 tons of biomass per hour. The project called CHRISGAS has the aim to produce hydrogen-rich gases from biomass, including residues without using the more costly pyrolysis path. One of the key elements of the CHRISGAS process is the use of a very high gas filtration temperature in the range between 800 and 900°C. This gas can then be upgraded to commercial quality hydrogen or to synthesis gas for further upgrading.

Thermal gasification initially converts biomass into a synthesis gas, a gas mixture that contains varying amounts of carbon monoxide and hydrogen not unlike the town gas we discussed earlier.

As regards commercialisation of hydrogen from biomass, a progressive company in Brazil, RAUDI is basing its work on gasification of sugar cane. Their work aims at developing turnkey plants ready for the market within a few years. In Brazil, some 215 million tons of dry biomass would be available for this process, leading to about 7 million tons of H_2. RAUDI quotes the target cost of 1.87 USD/kg H_2.

In the gasification process an outlet can be made at various points in the material flow in order to utilise the product already as fuel. For example it is possible to produce a dimethylether DME which in Sweden been used successfully as a liquid transport fuel for diesel engines. In Japan a 100 ton/day demonstration facility has been operating supported by Ministry of Energy, Trade and Industry.

Another possibility is to convert the biomass to Hythane (Registered trademark) where hydrogen is mixed with methane. By a mixture of up to 20 per cent H_2 in CNG the combustion in the cylinder will be speeded up and will reduce emission of methane. This has been demonstrated in for example Sweden and Canada in compressed natural gas (CNG) vehicles.

In the paper and pulp industry there is an interesting and exotic source of hydrocarbons to mention, the so called residual black liquor which contains mainly carbon (38 per cent) with sodium (17 per cent), some sulfur and oxygen similar to the carbon content, and finally a small (4 per cent) amount of hydrogen. Black liquor can be used in a gasification process or pyrolysis, but is also utilised by combustion. The input energy exceeding 500 MW for a typical paper and pulp plant can yield considerable energy in the form of black liquor.

Combination of gasification with modern membrane technology is a worthy research subject. The Gas Technology Institute of the USA has been testing proton conducting membrane materials development for optimum hydrogen flux. For example membranes made of Barium Cerate Perovskites, the basic crystal structures of which are quite common in minerals.

Biological methods of producing hydrogen

Biological hydrogen may be produced by biophotolysis of water by cyanobacteria and algae, by photodecomposition of organic compounds with photosynthetic bacteria and by fermentation from organic compounds. Figure 22 shows some of the most common pathways.

Fermentation takes place in the dark involving for example clostridias and thermophilic anaerobic bacteria. Mother Nature has a way to speed these processes with the aid of hydrogen-producing enzymes, of which Fe-hydrogenase, nitrogenase and NiFe hydrogenase are the most common ones. These enzymes exist in some microorganisms. The use of thermophilic bacteria and hydrogenase for hydrogen production enjoys considerable interest in the research community.

The hydrogen produced originates in water. The by-products are CO_2 as well as for example acetic acid. The process unfortunately becomes thermodynamically less favourable as the hydrogen concentration increases. Other microorganisms quickly consume the hydrogen produced. Hence, the composition of microflora in the reactor has to be limited to the hydrogen-producing varieties. A good reactor must take care of continuously removing the produced hydrogen from the process.

Planet Hydrogen

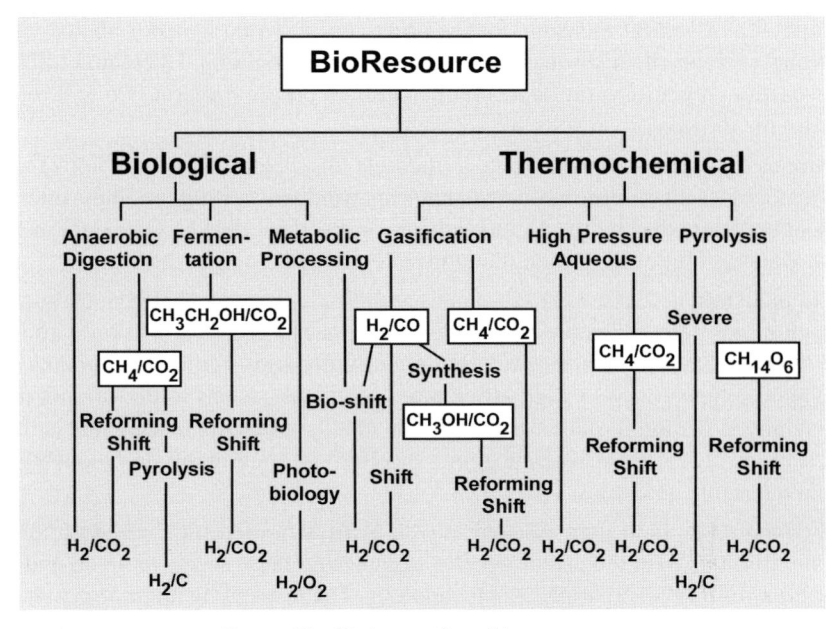

Figure 22. Hydrogen from bio resources.

The acetic acid from the dark fermentation process can be treated with light by the use of for example purple bacteria and hence be converted to more hydrogen.

The main hurdle in adopting these natural processes involving light is the relatively low energy conversion efficiencies. In other words: a small proportion, perhaps as low as 1 per cent of the power of the light is effectively converted into the hydrogen. Furthermore, the smallest presence of oxygen in the otherwise oxygen deprived environment inhibits the productivity of the bugs, even at oxygen concentrations as low as 0.1 per cent their hydrogenase production is severely hampered. Light saturation effects have also been observed as well as efficiency losses when the microorganisms start converting a part of the light directly into heat.

An exciting future development is the road involving photobiological processes such as photosynthesis in cyanobacteria or algae. Biological electrolysis (splitting water into hydrogen and oxygen) is the first step in photosynthesis.

Photosynthesis occurs when chlorophyll in green plants or green and blue-green algae absorb sunlight. Enzymes use this energy to break water down into hydrogen and oxygen. The hydrogen is subsequently combined with carbon dioxide to form carbohydrate. Certain microorganisms release hydrogen instead of carbohydrate during photosynthesis.

In many ways such methods are imitating the fundamental processes involved in the birth of oil and hydrocarbons in nature – just made faster by controllable means.

There exist enormous challenges for research in this exciting area of nature imitation! This part of the science is called *biomimetic hydrogen* production. Nature uses chlorophyll as a central molecule for the process of utilising light for photosynthesis. The Nobel prize in chemistry in 1988 was given to Deisenhofer, Huber and Michel for the discovery of photosynthesis reaction centres in purple bacteria. We will return to this after the discussion on solar hydrogen.

Green algae and cyanobacteria (remember the beginning of the book) are the main categories of microorganisms that can produce hydrogen. They operate at different wavelengths and the photosynthesis of cyanobacteria can use light in the infrared, whereas algae can go into the ultraviolet part of the light spectrum.

In industrial processes, the organisms are initially kept incubated in darkness and deprived of oxygen (anaerobic conditions). Once they are exposed to light after their prolonged adaptation to life in the dark, they initially emit significant amounts of hydrogen and oxygen (at very high efficiencies) until they revert to performing normal photosynthetic carbon dioxide reduction. The challenge facing researchers is that the algae saturate at low solar irradiances and thereby lose their ability in producing commercially useable quantities.

Fermentation, both light assisted as well as dark fermentation, enjoys a lot of attention currently. With the use of genetic engineering there seems to be no end to the possibilities of this "in principle" old technology. The fermenting microorganisms have an inherent weakness to oxygen and the element acts like poison in their primitive life. Hence, research is aimed at increasing the enzymes´ tolerance to oxygen. Perhaps there is a revolutionary ground breaking advanced species of microorganisms, anthropogenic – designed by genetic engineering, or natural, lurking out there awaiting to be put into action. They still remain to be found.

Hydrogen from nuclear energy

When examining general thermochemical hydrogen production apart from biomass, we must finally mention the sulfur-iodine process as an example. This process, which concerns decomposition of water, was originally developed by General Atomics in the 1970s and was intended mainly for the nuclear power industry. Hydrogen from nuclear power can either be made by electrolysis which we discuss in the next section or by thermochemical methods.

The nuclear industry is performing experiments with the so called Very High Temperature Reactor concept utilizing a graphite-moderated core with a once-through uranium fuel cycle. This reactor design envisions an outlet temperature of 1,000 °C. The high temperatures enable applications such as process heat or hydrogen production via the thermo-chemical sulfur-iodine process. The thermal efficiency is expected to be very high.

The sulfur-iodine process involves iodine and sulphur dioxide, emits oxygen but remains a closed cycle. Many of these processes are so corrosive that they are often termed as "plumber´s nightmare" and demand a lot of detailed knowledge about materials degradation and strength.

Another important method for reducing water at high temperatures is the Bromine/Calcium/Iron process, sometimes called UT-3. It consists of chemically treating water by compounds of the three elements in a four step process that results in hydro-

Planet Hydrogen 61

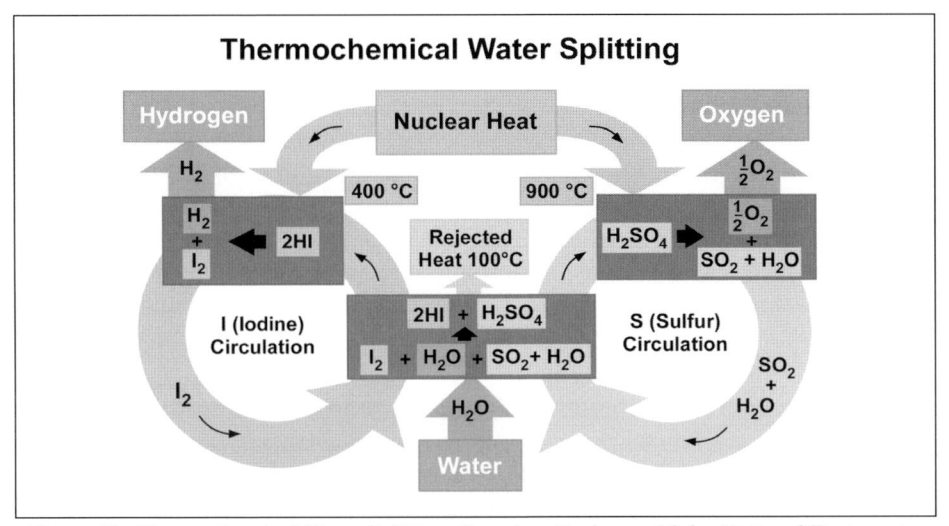

Figure 23. Thermochemical Water Splitting. Based on Ryskamp, Idaho National Engineering and Environment Lab.

gen release. This process needs slightly lower temperature than the sulphur-iodine method. (Figure 23).

The Sulfur-Iodine process and the UT-3 are described in Cyber-Appendix III where advanced thermochemical cycles are shown.

"Jules Verne´s method"

When Jules Verne wrote his famous novel on the Mysterious Island which was published in Paris in 1874, and prophesised "water replacing coal", he was well aware of the technology of *electrolysis*, which was discovered by William Nicholson, 1800. The water molecule is held together through strong molecular forces. These ties need to be "broken" by the use of electric energy. Electrolysis is a word from Greek with "lysis" standing for "cutting" or separating. Breaking the bonds requires to separate water into its elemental constituents hydrogen and oxygen. The chemical reactions are the following:

At the cathode, the negative electrode, the electron attacks the water molecule to form hydrogen

$$2H_2O + 2e^- \Rightarrow H_2 + 2OH^-$$

At the anode, the positive electrode, the oxygen is delivered

$$H_2O \Rightarrow 1/2\, O_2 + 2H^+ + 2e^-$$

And the combined reaction where energy and water become hydrogen and oxygen:

$$33.6\,kWh + H_2O \Rightarrow H_2 + 1/2\,O_2$$

When all ingredients are gaseous, this energy amounts to about 33.6 kWh per kilogram of steam. We only need to keep in mind that this figure will be higher for liquid water because the energy to convert water to steam is considerable. The figure we have obtained is usually called the lower heating value (LHV) which is lower than the higher heating value (HHV) of 39.7 kWh. This can be very important in any comparisons of efficiencies etc. Figure 24 shows the hydrogen refueling station in Iceland, which uses on-site electrolysis for the production.

Figure 24. On site electrolysis and storage in the hydrogen re-fueling station in Iceland.

Compared to fossil fuels such as petrol and diesel, a given mass of hydrogen contains about three times more energy.

Industrial electrolysis is well known in many countries and territories where available electricity is ample. These are countries like Brazil, Canada, Egypt, Iceland, New Zealand, Norway, the island of Tasmania and Zaire. In France, Belgium and Switzerland where nuclear energy is an important part of the energy mix, electrolysis is also a potentially important method. It need be kept in mind, however, that the amount of hydrogen produced by electrolysis in the world is very small compared to fossil fuel based hydrogen, or about one part in a thousand or so. The most common single electrolysis production process for hydrogen is the chlor-alkali process where the element is a by-product.

When water is split by electrolysis, the water itself serves as the electrolyte separating two electrodes, metallic entities that carry the current from the electric source needed to break up the bonds. The water has been mixed with an alkaline compound to make it more conductive. The direct current voltage, theoretically 1.481 Volts at room temperature, gives each of the two electrons needed per water molecule, sufficient energy to split the water molecule. In industrial facilities the voltage is typically from 1.7-2.1 V. The amount of water needed for electrolysis is about .8 litres per normal cubic meter of hydrogen obtained. The hydrogen is released at the negative electrode (cathode) and oxygen at the positive electrode (anode) as shown in the chemical equations earlier. The ions of the electrolyte, the hydroxide having been enhanced by the alkali compounds, and protons of water adjust the charge balance of the bath. The anode and the cathode are separated by a microporous diaphragm to prevent the mixing of the product gases.

The current research into classical alkaline electrolysis addresses the use of high pressures and more economical materials. One of the consequences of going to higher pressures is to avoid gas bubble formation. Usually, the gas formed at the electrode has a tendency to form bubbles with the resulting poor penetration of electric current. In the case of higher pressure bubbles are less of an obstacle. Bubble formation can also be prevented by proper electrolyte circulation. The main advantage of applying higher pressures is that the internal electrical resistance of the electrolyser is lowered so that the overall energy efficiency is increased.

When we study research activity in this field we note that for example tests have been made by using temperatures up to 300-600°C and electrolysis of the water vapour using a hydroxide salt melt electrolyte. Many corporations and institutions have been contributing in this area. The challenges are for example to reduce resistive losses of the current during electrolysis. Wherever electric current passes through matter it will be met by resistive response. In fact we utilise this for heating in our electric water kettle. Such losses will in general increase the voltage across the electrolysis path proportionally to the electrolysis current and resistance developed. This is most often called IR losses. They can be reduced if the temperature is raised but only up to 150 °C when working with the most common electrolytes NaOH and KOH.

A PEM electrolyser, where the electrolyte is a solid substance, will be discussed in the chapter on fuel cells later in the book.

A number of companies around the world produce electrolysers and are constantly improving their quality and durability. One of the new developments, besides the PEM electrolyser, is the high pressure electrolysers that can go up to pressures around 120 bar. By going to higher pressure the producer kills two birds with one stone: increased efficiency is obtained and not less important, the need for special compression unit for storage is reduced or eliminated.

Finally, we will comment on the use of higher temperature to electrolyse water. In fact, the electric energy needed for electrolysis is reduced if the temperature of the bath is increased. There have been experiments with high temperature electrolysis since the last decades of the twentieth century with some promises..

High Temperature Electrolysis (HTE) has been tested since the early 1980s in Germany by Dönitz and his coworkers. At that time, the process which was nicknamed HOT ELLY, was deemed not sufficiently economical to be pursued further. In the range of 800-1000 °C the required electrical energy for electrolysis is reduced by roughly 25 per cent compared to conventional electrolysis. Also HTE allows for more compactness and the use of higher current densities. SOFC fuel cells which we will discuss at length in a later section can be used in a reverse mode to electrolyse water. The outcome, the solid oxide electrochemical converter, SOEC has been shown to have efficiencies close to 92 per cent, almost 50 per cent more efficient than the traditional alkaline electrolysers. With SOEC, Mogensen and his fellow researchers at Risö Laboratory in Denmark have made remarkable progress using this method and slightly altered electrodes.

The high temperature solid oxide electrolysers have the advantage of being able to split CO_2 into CO and O_2. A mixture of CO_2 and steam can be electrolysed to give a mixture of H_2 and CO, a synthesis gas, from which methane CH_4 and methanol CH_3OH can be easily obtained. Of course methane made this way would be regarded as involving reusable CO_2.

The nuclear research establishment of CEA in France has done very interesting work on HTE. The French CEA combines HTE with their new generation of high temperature nuclear reactor. Supplying the heat for HTE from outside has coined the name *allothermal* for this process. Another route, devised by CEA, assumes heat provided by, for example, geothermal heat source and electricity from a hydroelectric facility. Steam will, in this case, enter the electrolyser and be heated by simple Joule heating provided by the electric current in the electrolyser and its ohmic resistance. This is called the autothermal route. CEA, the University of Iceland and more partners have been looking at this possibility under in what is called the Jules Verne Project using geothermal heat before finally converting to nuclear heat.

Hydrogen from wind

The harnessing of wind energy suffers from the limitations set by natural temperaments of the blowing winds. In a large grid connected system, the grid can always absorb extra energy and provide electricity in conditions of no wind. One possible use of electrolytic hydrogen is to store wind power to even out the fluctuations in production due to natural variation of the wind.

On the island of Utsira in the Atlantic Ocean just outside the magnificent fjords of Norway close to Haugesund and the ancient Viking time heritage, Norsk Hydro and its partners have been running a wind-generator based hydrogen system since 2004. Two wind generators of 600 kW each deliver electric energy to a small electrolyser which

Figure 25. The wind based hydrogen station at Utsira, Norway.

stores its hydrogen under pressure. The hydrogen is used for powering a number of houses on this small island. (See fig. 25). There is a 12 kW fuel cell from a Danish producer, IRD, delivering the power in addition to some 50 kW of electric energy stemming from an internal combustion-based generation unit. The combustion generator has an appreciably longer start up time than the fuel cell and the interplay of the two has proven a development challenge. The system also uses a flywheel for stabilisation. This first proper wind-hydrogen system in the world, is providing valuable experience for the coming generations of wind hydrogen systems.

Denmark gets more than 20 per cent of its electricity production from wind energy – with a resulting world leading know-how position. When wind is used as a secondary source with some more stable energy source as a part of a grid, the windpower has to be limited to a certain maximum fraction of the whole. The experts at Risö Institute near Roskilde have for a long time been experimenting with wind and diesel combinations. Their optimal recommended proportion of diesel in a joint system is somewhere near 20 per cent diesel against 80 per cent wind power. Small island communities could benefit from such a system and the experience in Utsira island in Norway could pave the way. Let us not forget that the problem of wind based energy shows up during days of calm weather. The Atlantic islands are not particularly known for still winds although such periods occur. The most important energy safety and security issue for such communities could be proper hydrogen storage.

Costs of wind-based hydrogen are rather difficult to pin down. The US Department of Energy has set a target for hydrogen from electrolysis, based on grid power amounting to 2.85 USD per kilogram of hydrogen produced.

Solar hydrogen

The use of solar energy to produce hydrogen is by many seen as the ultimate goal of renewable energy use. We refer to our discussion of solar energy harnessing in a previous chapter. In figure 26 we can see the various options open to solar hydrogen production. Solar-to-hydrogen conversion efficiencies can vary from as low as 1 per cent in the case of a natural photosynthesis conversion to biomass – up to a fancy 67 per cent for tandem, multi-photon and multi-bandgap photoelectrolysis devices which are being tested experimentally.

The solar-thermal electric power technologies used to operate advanced alkaline water electrolysis are the most efficient and commercially available methods today. Solar-to-hydrogen net efficiencies up to 20 per cent have been reported, using either Stirling engine or steam turbine systems based on the solar heating provided through a paraboloid dish. Another and quite different variant is the integration of a photovoltaic PV cell into monolithic devices with electrolysis. In such a system a PV cell operates with alkaline water electrolysis and is capable of reaching up to 16 per cent efficiency.

Solar powered Sulfur-Iodine or Bromine/Calcium/Iron processes as we discussed in the section on nuclear, and are explained in Cyber-Appendix III, are also possible and promising ways of utilising solar energy for hydrogen.

An exciting relatively new development is taking place in the field of photochemi-

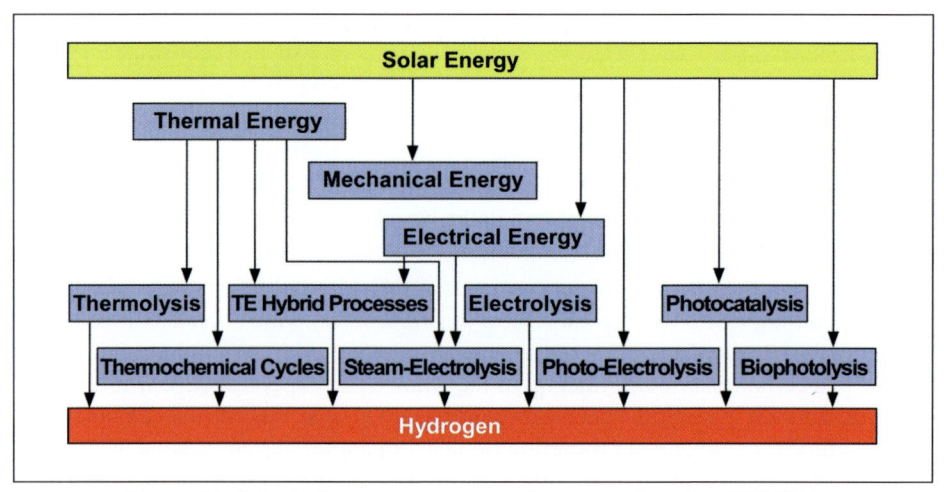

Figure 26. Solar hydrogen conversion paths options based on IEA-HIA.

cal hydrogen production which was first introduced by A.K. Fujishima and K. Honda in 1972. In a *photoelectrochemical* system (PEC) a semiconductor material which is photoactive is arranged to form a junction in contact with a liquid (or sometimes solid) electrolyte. When such a device is lit by sunlight, electron-hole pairs are formed at the junction. These light-induced electron-hole pairs in turn drive a chemical reduction and an oxidation in the electrolyte of the PEC system.

Appreciable research work has gone into characterising various materials for use in the PEC system. Of the various photoanode materials suggested we can mention tungsten trioxide, titanium dioxide, iron oxide, gallium arsenide, silver chloride and many more. The art of using these materials is to introduce controlled impurities which can work as a stepping stones for a jumping electron or hole inside the material. This is called doping.

The various projects with PEC which are run by researchers all around the world certainly are a tribute to the inventive spirit of modern science. Two kinds of systems are studied: Firstly, consisting of one photo-electrode and one metal electrode and secondly a more sophisticated system comprised of one n-doped in addition to one p-doped photoelectrode system. In this technology one sometimes gets the feeling that humans are approaching the creativity of Mother Nature – making artificial plant leaves. We refer to a picture of a water splitting system shown in figure 27.

Photobiological hydrogen research has evolved considerably during the past decades. Initially this work was based on screening and characterising hydrogen-producing micro organisms. The main constituents here are green algae and cyanobacteria. Genetic manipulation of the ability of these microorganisms to produce hydrogen is regarded as a promising way forward.

If we go from lower to higher temperature processes using sunlight we come to high temperature *solar thermochemical* production processes of hydrogen. They can involve five possible pathways: Thermolysis, thermochemical cycles, reforming, cracking and gasification. In four of these cycles, water is the main raw material to be used

Planet Hydrogen

PEC hydrogen production using a semiconductor photoelectrode

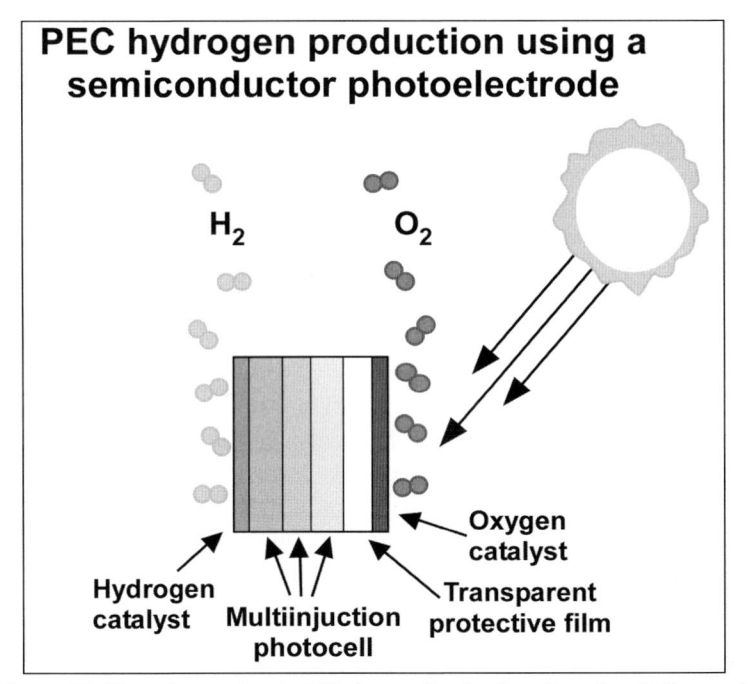

Figure 27. Photoelectrochemical Hydrogen Production. Based on Miller et al.

with the solar heat in a process called thermal dissociation; in cracking, however, fossil fuels are the major primary resource.

In the thermal dissociation zinc oxide is used for breaking up into zinc and oxygen. The zinc oxide consequently can split water.

In the cracking, which in fact is a thermal splitting of the long chain hydrocarbons into shorter ones, the carbon from the fossil resource is diverted into CO and CO_2, leaving hydrogen as a product.

A European project, HYDROSOL, has concentrated its efforts on making a catalytic monolith reactor based on solar heat for producing hydrogen at a very high efficiency of up to 80 per cent at 800°C. The basic idea is to combine a support structure capable of achieving high temperatures when heated by concentrated solar radiation, with a catalyst system suitable for the performing water dissociation and at the same time capable of regeneration at these temperatures, so that the water splitting and catalyst regeneration can be achieved by a single solar energy converter.

The overall solar-to-hydrogen efficiency is about 7-8 per cent. This project was awarded The Global 100 Eco-Tech Award at the Expo in Japan 2005 as an outstanding project paving the way for a sustainable future.

Mimicking the Maple Leaf

Chlorophyll is one of the wonders of nature in its use to harness the Sun's energy. Arranged as a receptor facing the sunlight chlorophyll, bound to an enzyme called "photosynthetic reaction centre", thus captures its energy and uses it for making chemical compounds.

A group at the University of California at Berkeley has been experimenting with trying to control the "antenna size" of the chlorophyll receptor in a given leaf of a tree or a microorganism. The group has identified a novel gene in Chlamydomanos reinhardtii assumed to be involved in the signal transduction pathway leading to regulating the chlorophyll antenna size in photosynthesis. Other groups are even mimicking the chlorophyll which we will take a closer look at.

In Uppsala in Sweden, where the Vikings assumed to be the dwelling place of the Nordic god Odin, there is a thriving university. A group of 40 researchers led by Professor Stenbjörn Styring has been working on artificial photosynthesis and made considerable progress in a very exciting area of new research in hydrogen.

The aim of the work is to create chemical catalysts for splitting water with the aid of light. To do this the Styring group looks back onto the way nature does this. Chlorophyll, the green pigment in the plant leaves, is found in membranes of chloroplasts within the leaves. The light reactions in plants are known to be highly efficient, or up to 40-50 per cent.

When light hits the chlorophyll, the electrons from the water molecule in the leaf move into an electron acceptor. Mother Nature uses manganese-based compound acting as a catalyst. The electrons end their journey by creating a carbohydrate with carbon dioxide as a building block. The electrons are extracted from water which is split into oxygen and hydrogen ions. The Styring group identified an amino acid called tyrosine as being a crucial link between the manganese complex and the chlorophyll. A key person in this was Ann Magnuson.

Instead of chlorophyll, Styring's group uses ruthenium based complex compound. Ruthenium is an element in the group of noble metals and has atomic weight 101. It is commonly found in platinum metal mines. The ruthenium complex behaves like chlorophyll in that it absorbs light of comparable wavelengths and has similar oxidation potential strong enough to affect the oxidation of water. The ruthenium complex seems more robust than chlorophyll which is remarkably sensitive to light. This

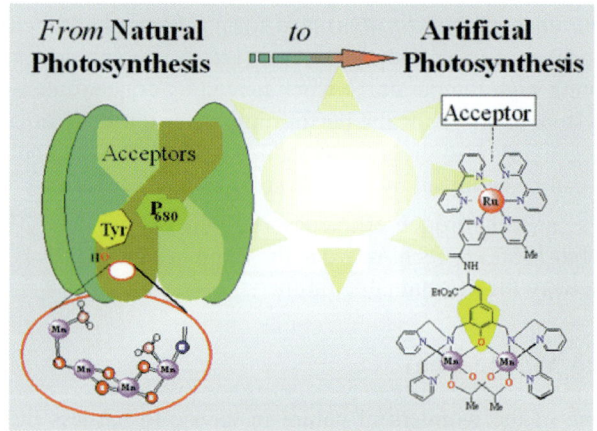

Figure 28. Biomimetic concept (picture from Styring group).

gives the group hopes that the artificial photosynthesis will be manageable and require relatively low maintenance. Similar to what we described for chlorophyll, the manganese complex extracts electrons from water and then in turn delivers them to the ruthenium. Nine years ago the group managed to link the ruthenium complex with the manganese complex to create a working electron movement bridge.

The remaining puzzle for the Styring team is to understand how, like in nature, the manganese complex can provide four electrons simultaneously in order to facilitate the full splitting of the water molecule. The group works day and night in striving to finally solve the remaining mystery of the "mimicking of the maple leaf" (Figure 28).

Hydrogen from geothermal vents

Our planet is glowing hot inside. The heat from the Earth's mantle continuously flows outward. It conducts to the surrounding layer of rock, the crust. When temperatures and pressures become high enough, some mantle or crustal rock melts, becoming magma. Then, because of its relative lightness compared with the surroundings, the magma rises, moving slowly up through the Earth's crust, carrying the heat from below. Sometimes this creates lava flows. Most often the magma leads to the heating of rock and groundwater. Some of the hot geothermal water and gases travel up through faults and cracks and they reach the surface as hot springs or geysers, but most stay deep underground, trapped in cracks and porous rock thus forming natural geothermal reservoirs. The geothermal gasses can contain a number of compounds, carbon oxides, sulfides, methane and hydrogen, to name a few. In a few cases the amount of molecular hydrogen is probably sufficient for economically feasible production.

Hydrogen sulfide is a common compound in many places of the world, bound in water/sea or in geological form. In the geothermal areas of Iceland considerable amount of H_2S is for example emitted to the atmosphere. In the already harnessed parts of the Icelandic high temperature regions well over a thousand metric tons of hydrogen could be recovered annually, thus turning a polluting effluent into renewable fuel. In geothermal boreholes in the Krafla region in Northern Iceland where the whole area is very close

Figure 29. Hydrogen from a geothermal borehole. Source: National Energy Authority, Iceland.

to a magma chamber and hydrogen is probably produced by some sort of a steam-iron-like process (the old method of developing hydrogen by flushing a glowing hot iron surface with steam), there are boreholes emitting up to 50 metric tons of molecular hydrogen annually. The borehole gas is hydrogen mixed with hydrogen sulfide, H_2S, carbon dioxide, CO_2, nitrogen, N_2 and of course steam. (Fig. 29).

In a project undertaken at the University of Iceland by my group, hydrogen was recovered from the borehole gas using a closed cycle of acidic ferric chloride solution which the geothermal gas is bubbled through. The ferric ions are reduced to ferrous ions and the sulfide ions become oxidised to solid sulfur that can be separated from the solution. When the remaining ferrous and hydrochloric acid solution is subjected to electrolysis, the ferrous ions are oxidized back to ferric ions at the anode and hydrogen ions reduced to recoverable hydrogen gas at the cathode. The electric energy needed to split hydrogen sulfide into its components has proven to be 2/3 of the energy needed for splitting water by electrolysis. The electric energy for the electrolysis is obtained at the power plant by flash turbine generators. Solid sulfur is the main byproduct of the total process.

The geothermal vents are indeed an interesting field of utilisation but there is another even more important domain awaiting larger scale production. An example of a

Figure 30. The Black Sea – huge reservoir of H_2S.

territory enormously rich in hydrogen sulfide is the *Black Sea,* which really is a living tribute to an ancient fossil producing seabed like the one which fostered the fossil deposits on Earth in the dawn of oil formation. The Black Sea (fig. 30) is literally infested with H_2S. Below about 170 metres of depth its waters contain no oxygen and very limited life in a vast hydrogen sulfide dominated world. The maximum depth is over 2200 m. A concentration of 7 mg/l of undissociated hydrogen sulfide has been reported from an anoxic region of the Black Sea at depths of around 200 m - making it the largest anoxic basin of the global ocean. About 4.6 billion tons of H_2S are estimated to be contained in the Black Sea. The name of the sea, given by seafarers, is believed to originate from the inhospitable environment sometimes caused by storms rather than its murky abyss.

STORING HYDROGEN

Introduction

Devising an effective form of storage is the main challenge for the use of hydrogen in society. Presently, hydrogen is being stored in either gaseous, liquid or solid forms. Each method has distinct pros and cons. Most of the 50 million tons produced world wide annually is stored in the form of gas. Hydrogen storage has been called the greatest bottleneck for development of the hydrogen economy. In fact, none of the existing methods for storing hydrogen are sufficient, neither on a volumetric nor per unit mass basis. Compared to fossil-fuels like diesel or petrol, hydrogen suffers from its large volume and/or significant weight of storage containers. Figure 31 shows the volume of 4 kg of hydrogen stored in various different forms relative to the size of a typical car.

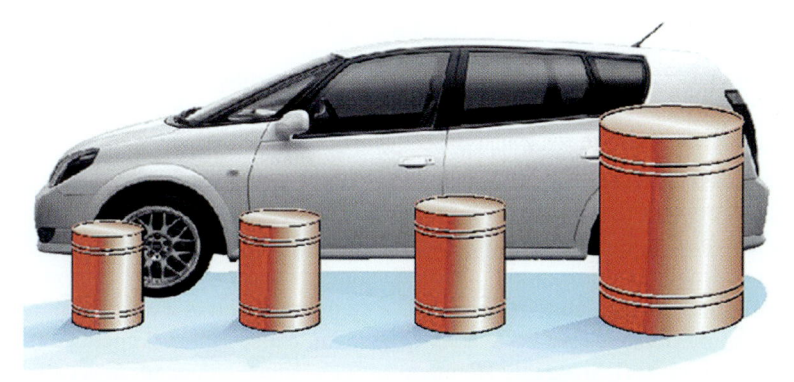

Figure 31. 4 kg of hydrogen relative to the size of a car. Different storage methods are shown from left to right: Mg_2NH_4; $LaNi_5H_6$; $H_2(liq.)$; $H_2(gas\ at\ 200\ bar)$. Based on Schlapbach and Zuettel.

The US Department of Energy (DoE), in their challenges to the research and development community has set a storage target of 6.5 per cent hydrogen by total weight and 62 kgH_2/m^3 for volume. The former figure means that in order to store 4 kg of hydrogen and meet the DoE target, the mass of the hydrogen "container" must not exceed 61.5 kg.

Storage in the form of gas

The most common method of storing hydrogen in automobiles at the onset of the 21st century is compressed gas at 350 bar (equivalent to 5000 psi or 35 MPa) pressure. The CUTE/ECTOS hydrogen-drive busses in Europe store about 40 kg on a full tank. This amount of gas is stored overhead in steel cylinders covering close to a third of the roof area of the bus. Compressed hydrogen tanks rated for 350 and 700 bar pressure have been certified worldwide according to ISO 11439 (Europe), NGV-2 (U.S.), and approved by TÜV (Technischer Überwachungsverein Germany) and The High-Pressure Gas Safety Institute of Japan (KHK).

Generally, larger volumes of hydrogen are stored at much lower pressures. As an example, the largest container in the world stores about 15,000 cubic metres of hydrogen, but uses pressure of only 12-16 bar, i.e. twelve to sixteen times atmospheric pressure.

The use of high pressure for storing hydrogen has decades of tradition in industry. In the central industrial Ruhr region of Germany there are hundreds of kilometres of hydrogen gas pipes as a part of the underlying infrastructure. The overall hydrogen pipeline length is 220 km; the pipe being 100-300 mm in diameter with a typical pressure of 25 bar. The pipe system connects 4 producers with 13 companies with a total hydrogen consumption exceeding one hundred thousand tons annually. Conventional mild steel has been used in Germany and France since 1938 as pipeline material, but new materials are being introduced as we will come to later.

In the southern United States a 400 km hydrogen pipeline services close to 50 customers in the area surrounding Lake Charles in Louisiana, Houston and Texas City, Texas.

By using fibre reinforced composites, companies like Dynetek of Canada have developed hydrogen cylinders able to operate at 825 bar for stationary applications and providing 700 bar storage for transport.

Compressing hydrogen is generally much cheaper than liquefaction. As an example, we can consider that compressing hydrogen from atmospheric pressure to 450 bar requires less than a tenth of the energy needed to produce it by electrolysis. Keep in mind that hydrogen compression is a very demanding task because of compressor seals and exotic materials needed to achieve leak free operation. It is well known that due to its tiny size and diffusive properties, hydrogen can cause embrittlement of metals. In the case of carbon steel; at elevated temperatures hydrogen can result in decarburization and a subsequent deterioration of equipment. When decarburization occurs, the carbon atoms intended to stabilize the steel matrix, are removed due to the formation of methane molecules within the metallic crystal. Alloy steels containing chromium and molybdenum have been suggested for compressor materials that are resistant to this process.

Cavities in salt deposits, usually formed by flowing water, can be converted into large-scale storage of hydrogen as well as natural gas. In the Tees Valley of northeast England where many large salt caverns sit abandoned, a salt dome is used to store approximately one thousand tons of hydrogen for industrial use. The experience from storing natural gas in this way is benefiting hydrogen technology. Additionally, the Tees Valley features hydrogen pipelines up to 30 km long. The aim of the Tees Valley project has been to help bring new energy technologies from development to the marketplace. The unique asset of the dome, together with the strong technological base and experience in hydrogen handling has spurred the creation of this project named after the valley. Rock cavities are possible but less likely to be an option for hydrogen storage mainly because their excavation is more expensive.

A long tradition exists for storing hydrogen in metallic cylinders of various sizes. The most common metallicl cylinders are made of stainless steel with added chromium, nickel or molybdenum. These alloys are titled AISI 304 or AISI 316 by the international characterization system.

A more modern design is based on composites, for example carbon fibre reinforced composites with aramide fibre and epoxy resin. Composite, 750 bar tanks have in tests demonstrated 2.35 fold burst pressure as required by the European Integrated Hydrogen Project specifications. In the U.S., bullet-proof outside reinforcement is further required.

Most hydrogen compression is done by cylinder-based compressors. The heart of a hydrogen compressor is often a hydraulically driven intensifier which consists of a motive cylinder, coupled with two gas cylinders. The most common examples on the market are either single or double stage compressors. The main disadvantage concerns the use of lubricating oils in these compressors potentially leading to oil contamination of the stored hydrogen and introduction of other impurities. We will discuss the potential of using metal hydrides for pistonless compression in the chapter on solid state storage.

Liquefaction and liquid storage

Hydrogen has the lowest atomic weight of all elemental species, and retains much of this quality when it forms a liquid of notably low density. Its density in the liquid form is only 70.8 kg/m^3, or to put into perspective, only seven percent that of water. Hydrogen cooling and liquefaction requires reducing the entropy of molecular states found in the system. Within the language of physics, cooling hydrogen gas can be described as *entropy reduction.* Likewise, in the words of F. Simon, a famous pioneer in refrigeration, a refrigerator is nothing more than an "entropy-squeezer". Squeezing is possible because for a given system at a constant volume, decreased temperature is equivalent to a reduction of entropy for the system. So for a liquid hydrogen bath with a small amount of hydrogen gas in the chamber above the liquid, the temperature of the bath can be lowered by decreasing the gas pressure in the chamber. Over a century ago, James Dewar succeeded in the first liquefaction of hydrogen. Hydrogen liquefies at about minus 253°C, hence special equipment is required for its storage. Two main challenges of liquid hydrogen storage are *energy efficient* liquefaction and *thermal insulation* of the storage vessel to limit boil off. The energy requirements for hydrogen liquefaction are high; typically 30 per cent of the heat content of hydrogen is required for liquefaction

The oldest hydrogen liquefiers were usually of the Linde type, using liquid air or liquid nitrogen as a pre-coolant. For the final stage of liquefaction, the hydrogen gas is made to expand in a so called Joule-Thomson expansion. During this process the gas undergoes a continuous throttling and expansion as it is driven by constant pressure on one side of the expansion valve (or porous plug) and expands to a lower pressure on the other side. In the particular case of hydrogen, there is one more complication involved: The hydrogen molecules can be of two different types, called *ortho- hydrogen* or *para-hydrogen* as we discussed earlier. The proton of the hydrogen nucleus has a magnetic property named *spin*. Spins can have two states: spin-up and spin-down. One can imagine the two protons in a hydrogen molecule having different spins and for ortho-hydrogen the spins would be parallel, whereas for para-hydrogen the spins would be antiparallel. As a result of fundamental physical laws, the two different states have very different energy content. This energy will show up in the value of the latent heat, the heat that must be extracted when lowering the temperature. The heat of conversion from ortho to para-hydrogen is 670 J/g, a little larger than the heat of liquefaction which is 450 J/g. As the temperature is raised, the fraction of ortho-hydrogen increases. At around 20K the para-hydrogen is completely dominant, whereas at liquid nitrogen temperatures, at about 77 K, the two states are almost equally probable. At room temperature the fraction of para-hydrogen is only 25 per cent.

Liquid hydrogen is usually transported by tanker trucks carrying about 4 metric tons of hydrogen with a loss of less than 300 kg during a typical transfer. The pressure of the tank gas is usually kept below 3 bars and the naturally evaporated hydrogen is vented out. Much care is taken to ensure that no air enters the liquid hydrogen tanks, flushing them with nitrogen gas is a typical procedure. In North America alone there are 10 hydrogen liquefaction plants with capacities up to 32 metric tons per day. The liquefaction process is very energy intensive and typical energy for large units are of the order of 12.5-15 kWh per kg hydrogen.

The liquid hydrogen infrastructure is small-scale and tends to service specialized projects. The National Aeronautics and Space Administration (NASA) utilises what is in all likelihood the longest liquid hydrogen pipeline in the world, a 600 meter installation in Florida. NASA has a total of 3,200 m^3 liquid storage tanks to ensure a continuous supply to the space programmes.

One can assume that the above liquefaction costs present a certain limit for the marketability of liquid hydrogen. In order to lower this limit, it is regarded as necessary to increase plant sizes and pursue high-speed centrifugal compressors and expansion equipment. These are just some of the tasks facing the development of hydrogen liquefaction. Moreover, new ways of liquefaction are presently being examined. These include *magnetic refrigeration,* where there are no pistons – no moving parts. A sequential magnetization and demagnetization of hydrogen gas by paramagnetic salts is used for the desired 'entropy squeezing'(remember what we discussed earlier) and liquefaction. Paramagnetic salts operate by lowering entropy via rearrangement of the magnetic moments of their atoms in such a way as to seek ever more order– resulting in decreasing temperature.

Thermal insulation is also a matter for constant improvement. This has been a challenge for the BMW motor company who have taken pride in running liquid hydrogen based cars with internal combustion engines. The BMW tanks usually can store about 120 litres of cryogenic hydrogen. A typical tank wall is about 3cm thick with 70 layers of aluminium foil woven into fibre-glass matting.

Leakages of hydrogen in long term parking of automobiles with liquid hydrogen storage amount to about 1 per cent per day and continue to be a technical challenge for this form of automotive application of hydrogen storage. Innovative solutions in reusing the lost hydrogen and utilising its energy have been presented. As of today BMWs tanks allow for an inner pressure of up to 3 bar prior to boil-off. Further, work is under way to utilise the boil-off in a small (10 kW) fuel cell (replacing the traditional electrical generator).

Solid state storage

While working with gas and liquid storage systems we quickly found out that hydrogen presented quite a challenge. Generally, for storing hydrogen we would like to pack it as close as possible; use as little additional material as possible and reduce the large natural volume demanded by the lightweight element.

A kilogramme of hydrogen in the gaseous state at standard temperature and pressure requires about 11 cubic metres of volume. We seem to have a few options: we can apply work to compress it, as we saw in the chapter on gaseous storage; we can "squeeze the entropy", lower the temperature and store the gas in the form of liquid. Alternatively, we can reduce the repulsion by letting the H-atoms interact with another material.

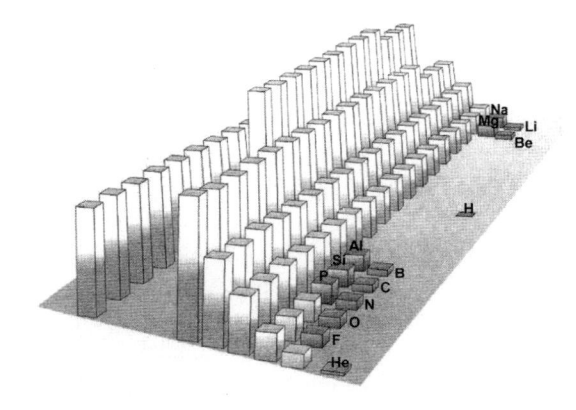

Figure 32. The light periodic table. The demand for light mass of metal hydride storage usually calls for elements from the low mass part of the periodic table.

We recall from the studies of the phase diagram that converting hydrogen to solid state requires enormous pressures and is outside our technical reach at present. This means that we have finally arrived at the important choice of storing hydrogen in chemical compounds or at least bound to different substances in the solid or liquid state. This is of course the way most hydrogen on Earth is stored in nature. Storing it in the form of water, methane or other hydrocarbons is not the option we will study now. Instead, we will turn our attention to the ability of materials to absorb hydrogen, alloy with it or at least compound with it in some shape or form – preferably alloyed with a light element. (fig 32).

Quite a few simple hydrides require less volume to store a kg of H_2 than liquid hydrogen. For liquid hydrogen in state-of-the-art containers, the average volume for one kg hydrogen is about 14 litres. For metal hydrides like titanium iron hydride (TiFe) the volume required is comparatively only 9.8 litres per kg hydrogen, while for lithium aluminium hydride the volume is 10 litres per kg hydrogen. Given these promising figures, let us look closer at what issues arise when considering the ability of materials to absorb and release hydrogen.

One of the elegant aspects of solid state storage of hydrogen compared to standard high pressure gas cylinders is that the solid storage offers design flexibility in terms of the shape of a container and its location – for example under an automobile. Application specific factors to keep in mind regarding hydride storage are: reversibility, light weight, rapid kinetics and equilibrium properties of pressure and temperature consistent with near ambient conditions.

In the rest of this section on storage of hydrogen we will elaborate on general aspects of solid state storage.

Sorption properties

Hydrogen can be absorbed into pre-existing solid structures through different mechanisms. The two primary sorption mechanisms commonly referenced in the field of solid hydrogen storage are known as physiosorption and chemisorption. Physiosorption can be thought of as (molecular) hydrogen gas entering into solution inside of a solid structure. This most commonly occurs at the surface of a material and can eventually reach a saturation point. Chemisorption is clarified as adsorption of a gas (i.e. hydrogen) into a solid lattice and subsequent chemical bonding and formulaic change of the material. The concept of physiosorption is important to hydrogen storage because it offers a method to store with reliable reversibility. Depending upon the ambient conditions and the porous nature of the adsorbent material, physiosorption can take several forms. The simplest mechanism is adsorption onto a surface, or monolayer adsorption. The monolayer adsorption capacity of hydrogen is a function of the specific material's surface area. Because molecular absorption on a surface is governed by weak so called Van der Waals-type forces, the "solubility" or adsorption capacity will decrease at higher temperatures as the average kinetic energy of the hydrogen molecules increases and they have a tendency to break away. Van der Waals forces are electrostatic in their nature.

Storage of hydrogen through physiosorption/adsorption has been studied exten-

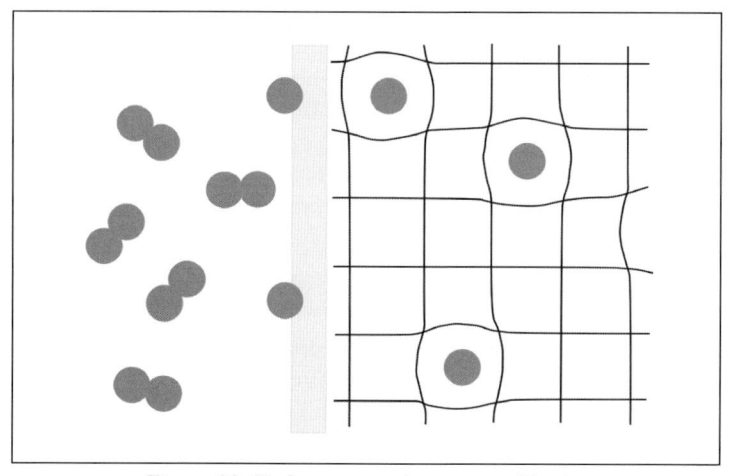

Figure 33. Hydrogen entering a crystal lattice.

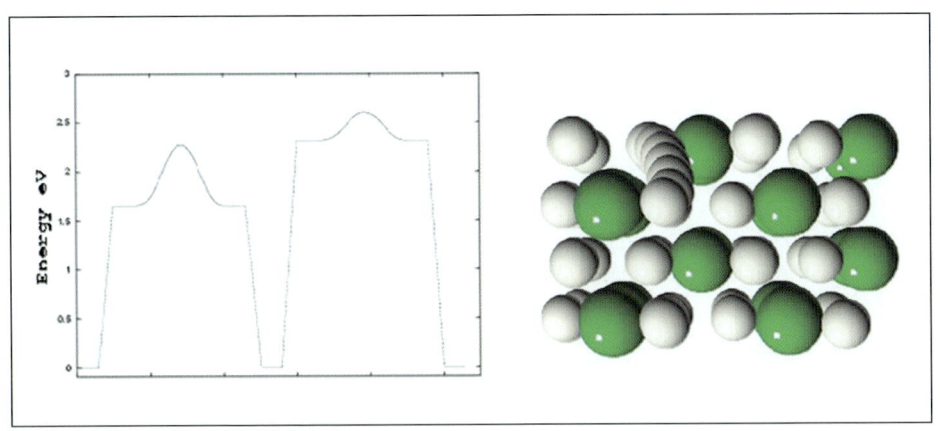

Fig. 34. The lattice of a metal crystal as it looks to a hydrogen atom looking for a nice valley to rest in and become metal hydride. Source: Hannes Jonsson, Iceland.

sively on carbon surfaces. If we consider that a monolayer of hydrogen molecules contains approximately 1.3×10^{-6} mol/m^2, then a graphene sheet with a surface area of 1315 m^2/g has the ability to adsorb up to 0.4 hydrogen atoms per each carbon atom. From that we can then calculate a weight capacity of 3.3 per cent hydrogen in a one sided sheet.

Other forms of carbon display quite different properties; activated carbon has been shown to adsorb on the order of two per cent weight capacity. Nanostructured carbon, on the other hand, demonstrates a much higher ratio of hydrogen adsorbed per carbon atom (0.95), translating to a theoretical weight capacity of 7.4 per cent (80 per cent reversible at 600 K). Among the carbon systems with a high surface area are Fullerenes (sometimes nicknamed buckyballs) and fullerene-derived nanotubes. Nanotube hydrogen storage created a great deal of excitement during the mid-1990s, but has since been plagued with irreproducible results and what now appears to be an incorrect initial assessment of hydrogen storage capacity due to the presence of water vapour wrongly assumed to be hydrogen.

Heben and Zhang at National Renewable Energy Laboratory in Colorado have reported that $C_{48}B_{12}(ScH)_{12}$ can hypothetically bind up to 11 hydrogen atoms per scandium (Sc) metal atom, 10 of which absorb reversibly (about 9 wt. per cent) as molecules at room temperature. The subject has received a great deal of interest from the scientific community.

Chemisorption or chemical bonding is the primary mechanism for the next class of solid hydrogen storage materials we will turn to. These materials absorb hydrogen and form new substances, commonly known as metal *hydrides*.

A hydrogen atom which enters a solid crystalline hydride material experiences various attractions and repulsions within the lattice. (fig.34). This can look like a landscape of valleys and hills. These are manifested in attractive sites and repulsive sites. Increased temperature provides the atom with the ability to climb the hill and eventually allowing the atom to diffuse deep into the lattice and come to rest in a deep 'energy valley'. A given atom will remain in its interstitial site until excited again by increased temperature. The figure shows the virtual landscape seen by a wandering hydrogen atom in a lattice.

Many metals and alloys react with hydrogen to form metal hydrides according the reaction:

Me + $x/2$ H$_2$ = MeH$_x$

Many elements can play the part of the metal - Me. According to the number of hydrogen atoms - x, connected to each metal atom, the amount of hydrogen molecules necessary in this equation is x/2. We will study this in much greater detail in Cyber-Appendix VI

Some of the most well-studied single-metal (binary) hydrides are MgH$_2$ and PdH$_{0.6}$. The hydrogen per mass percentages of each are 7.6 and 0.6 wt. per cent respectively. However, MgH$_2$ only releases stored hydrogen in any significant quantity above 330 °C. In order to understand why high temperature environments are frequently necessary, let us look deeper into the behaviour of hydrogen in metals.

If we follow the movement of a hydrogen molecule as it approaches a metal we note that the first attractive interaction happens close to the surface of the new medium. The force experienced at this stage, usually referred to as the Van der Waals force is rather weak, typically one fifth to one tenth that of the final chemisorption energy involved in bonding hydrogen inside the storage medium. In this initial physiosorption process, the hydrogen molecule interacts with several atoms at the surface of the solid. The attractive interaction between the hydrogen molecule and the solid is typically expressed as inversely proportional to a 6th power of the interatomic distance. Doubling the distance would thus result in 64 times weaker force. Additionally there is a repulsive term in the Van der Waals binding, diminishing exponentially with increased distance. The balance of these two happens at distances amounting to 2-3 Angströms. (To give the reader a feeling for this length, his corresponds to the typical distance between sodium and chloride in the structure of table salt, NaCl).

As the hydrogen moves into the metal storage medium, the atom has to overcome an activation barrier, first for dissociation and then for formation of the hydrogen-metal bond. In our energy landscape picture, this can be compared to making a great leap over a wall. This is the process referred to as chemisorption. After dissociation on the metal surface, the hydrogen atoms generally diffuse rapidly through the bulk metal even at room temperature to form a metal-hydrogen solid solution commonly referred to as an a-phase metal hydride. At this stage, the hydrogen atoms can occupy interstitial sites in the crystal lattice forming a tetrahedron when surrounded by four metal atoms - or a octahedron when surrounded by six metal atoms. In general the dissolution of hydrogen atoms leads to an expansion of the metal lattice which can amount to 30 percent. The overall expansion of the lattice can amount to a few percent. (Exceptions occur in the case of some rare earth metals where electronic effects lead to lattice contraction).

Conversion of the saturated solution phase into hydride takes place at a constant pressure in accordance with the the so-called 'Gibbs´ phase rule'. At this stage in the storage process, the pressure forms a plateaux of constant value as the concentration of the occupying hydrogen atoms is increased. When complete conversion to the hydride phase has taken place, further dissolution of hydrogen follows increasing pressure. Plateux can vary in nature and multiple plateaus are observed in composite materials. We will return to this in more details in the discussion of van´t Hoff plots in Cyber-Appendix VI.

Binary systems and beyond

An important group of hydrides consists of binary alloys composed of two elements. Inter-metallic compounds (IMC) or transition metal hydrides, represent a class of hydrogen absorbing materials that are made up of two or more different transition metals. The vast majority of IMCs consist of two main parts, usually referred to as the A and B component. It is easier to understand why these two distinct parts are necessary and how they allow IMCs to function so effectively at room temperature (or other desired ranges) if we examine the bonding nature of these compounds. On their own, the A and B components differ in the stability of elemental hydrides they form. Generally A components form stable elemental hydrides versus B components which form unstable (overly weak-bonded) elemental hydrides.

Effective IMC hydrogen storage allows use of a combination of A and B parts to, in essence, "average out" the stability and bond-strength of the final material. Lanthanum is a prime candidate for an A-type hydride and Lanthanum Nickel hydrides are an excellent example of this class of IMC materials. A compound such as LaH_2 requires high temperatures to function reversibly due to the extremely stable La-H bond. $LaNi_5$, on the other hand, undergoes hydrogen sorption at close to room temperature. The B component, Ni, bonds much more weakly and does not form stable hydrides that can be utilised reliably under the conditions needed. In comparison to LaH_2, the heat of formation (ΔH) of $LaNi_5H_6$ is much lower, 30.1 kJ/mole H_2 compared to 209 kJ.

In the inter-metallic hydride (IMH), absorbed hydrogen bonds directly with the non-hydride forming component – B, contrary to component A as one might expect. The B-H bond in the IMH is indirectly strengthened by its proximity to the A atoms in the lattice. This fact ensures that both components, A and B are necessary to effective hydrogen storage alloys. In $LaNi_5H_6$ this "medium strength" Ni-H bond allows for the material's characteristic room temperature desorption range. The important ratio of A to B in inter-metallic metal hydrides will be discussed further in Cyber-Appendix VII. There we will also comment on the use of these hydrides in hydrogen compression equipment.

Finally, an important comment on nickel metal hydride batteries. In the development of batteries, the knowledge discussed above is utilised in a form of batteries where nickel metal hydride is the core compound. They use nickel hydroxide for the positive electrode and hydrogen absorbing alloys, capable of absorbing and releasing hydrogen at high density levels, for the negative electrode. Such batteries can be charged and discharged in excess of 500 cycles and have more than double the energy density of the more conventional Nickel-Cadmium batteries.

Figure 35 shows the principle of a NMH battery and schematically how the protons govern the discharge.

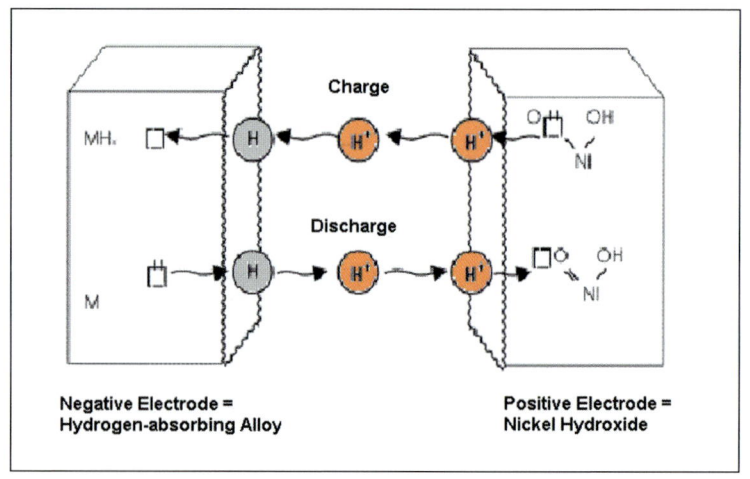

Figure 35   Principles of a Nickel Metal Hydride Battery.

Complex hydrides

The real wild card among solid hydrogen storage options is the *complex hydrides*. Complex metal hydrides represent a slight departure from the chemistry of the larger class of chemisorption-based storage materials. Upon the absorption of hydrogen, complex metal hydrides convert from metals (alloys) to either covalent or ionic bonded solids. This is in contrast to "conventional" hydrides which retain their metallic character. Overall, complex hydrides are not as well researched as other types of solid, gaseous or liquid storage and in a large number of cases, little is known at all. They are mainly attractive due to their high volumetric and gravimetric densities; some of the highest values reported are for Mg_2FeH_6 in which the volumetric hydrogen content is 150 kg/m^3. Similarly, in $LiBH_4$ the weight percent of hydrogen is enormously high at 18 per cent.

Alanates

An important issue that must be addressed in order to properly understand hydrogen storage properties of the previously mentioned compounds is the difference between thermodynamically limited and kinetically limited reactions. A good example of a thermodynamically allowed reaction, that is kinetically limited is the phase change from graphite to diamond.

The caveats with complex metal hydrides stem from their unique bond character. The transition from raw metal to a hydrogen containing compound is not a simple or direct process. In most cases there are numerous intermediate steps required to reach the final forms. This is the case of alanates, hydride compounds composed of alkali metals and aluminium. A popular example is sodium alanate, originally discovered by M. Schwickardi and B. Bogdanovic in 1997 at the Max-Planck Institute in Muehlheim an der Ruhr in Germany. Alanates have received a lot of attention by hydrogen storage experts. The basic compound $NaAlH_4$ releases its hydrogen in two main steps. In the first step, about half of the theoretical load of hydrogen is released. After the second step, up to $^3/_4$ of the total hydrogen capacity has been released. The chemical sequence can be simplified as:

$$6\ NaAlH_4 \leftrightarrow 2\ Na_3AlH_6 + 4\ Al + 6\ H_2 \leftrightarrow 6\ NaH + 6\ Al + 9\ H_2$$

Each of the intermediate steps in the liberation of hydrogen via decomposition of sodium alanate can be complicated and dependent upon a number of parametres as well as assisted by catalysts. The same applies to another well-known alanate, lithium alanate. Catalyst dopants operate by interacting with some of the electronic properties of the host alanate. The Max-Planck group used titanium to dope sodium aluminium hydride, resulting in about 3.7 wt. per cent hydrogen storage and desorption in a reversible manner. Dopants have displayed additional effects on alanates by lowering the desorption temperature.

In an interesting study by Peter Edwards at Oxford and Paul Anderson at Birmingham, UK, the generally sluggish hydrogen absorption/desorption behaviour in the powerful MgH_2 compound was improved by adding minute amounts of $LiBH_4$ to the compound and subsequently ball-milling the mixture. Ball-milling is a process which involves high velocity "supergrinding" with the aid of steel milling balls.

Eric Majzoub at Sandia Labs in the US and his colleagues have attempted to understand the mysterious role of titanium in accelerating the reaction above. Using Raman spectroscopy, using a laser beam for determining materials structure, they concluded that the titanium dopants lower the strength of the Al–H bond, conceivably flattening out the energy valley of the lattice landscape in the language we used earlier.

Current literature has reported the existence of around 70 complex metal hydrides alongside the alanate family. Another high hydrogen content family of complex metal hydrides constitutes the *borohydrides*. Because of the element boron's low molecular weight and its affinity to hydrogen, many of the borohydrides have storage capacity values that rival those of the alanates; the highest of these is in $Be(BH_4)_2$ at 20.8 wt per cent hydrogen. However, this compound is seldom considered for any practical purposes due to the toxicity of beryllium, (Be).

Borohydrides

The borohydrides have largely been neglected until recently when considering potential hydrogen storage materials. The borohydride systems manufacturer, Millennium Cell has stated that difficulty in direct combustion of borohydrides contributed to their initial abandonment shortly after their discovery as energetic compounds in the 1960s. Moreover, an another reason for their previous neglect is their purported lack of reversibility akin to that of the alanates. Borohydrides have yet to show the same responsiveness to the addition of foreign catalysts as the alanates do. Nevertheless, because of their chemical similarities, there is hope that a potential candidate may be found.

Millennium Cell has taken a different approach to the inherent problems in complex metal hydrides and borohydrides in particular. The firm developed a system dubbed

"Hydrogen on Demand" which utilises an aqueous solution of sodium borohydride (NaBH$_4$). The sodium borohydride solution is passed through a proprietary catalyst (onboard if in a mobile system) which hastens the pace of its reaction with water, providing the following products:

$$NaBH_4 + 2\ H_2O \longrightarrow NaBO_2 + 4\ H_2 + Heat$$
By catalyst

The reaction product, NaBO$_2$ or sodium metaborate is believed to be completely benign and at this point can either be discarded or collected for use in a recycling procedure. This is the main difference between the Millennium Cell design and that pursued by others researching similar compounds. The company chooses to forgo the requirement of reversibility for their concept of an effective solid hydrogen storage vessel. It is important to note that the raw materials in Millennium Cell's system cycle between common borax and the sodium metaborate byproduct, both of which are cheap by metal hydride standards. Additionally, Millennium Cell outlines a set of recycling reactions for converting the sodium metaborate to borax and then back again to sodium borohydride, however this cannot be performed onboard at this time. A prototype DaimlerChrysler vehicle called Natrium has been roadtested utilising a fuel cell system fed with hydrogen from Millennium Cell's sodiumborohydride system. The cost and composition of their catalyst is unknown but an educated guess would be that it involves supported ruthenium or some other transition metals.

Another interesting candidate is NH$_3$BH$_3$ - ammonia borane, also known as borazine. Borazine notable in its capacity to store close to 15 per cent wt. hydrogen, as reported by Wolf and colleagues at Technical University of Freiburg in Germany. The compound decomposes to polyaminoborane type compounds, which are toxic to both humans and most fuel cell components. Nonetheless, the high storage capacity has created interest in circumventing this hurdle, as seen by the efforts of Autrey, Gutowski and Gutowska at the Pacific Northwest National Laboratory in removing borazine from the hydrogen stream, by carrying out the reaction in a porous silica support.

Exotic Hydrogen Storage Compounds

As we approach the end of our discussion of solid or *condensed* hydrogen storage materials, it is of interest to mention a few exotic compounds. We can, for example, remind ourselves of Jules Verne's original idea about "water replacing coal". One interesting group of water-containing compounds is *clathrates* of which chlatrate hydrates are the most common. Clathrate hydrates constitute a class of solids in which the guest molecules occupy, fully or partially, a cage-like lattice formations in host structures made up of H-bonded water molecules.

In a remarkable work by Patchkovskii and Tse from Steacie Institute for Molecular Sciences, National Research Council of Ottawa, Canada, performed in 2003, this interesting problem is addressed. The medium researched is the interesting hydrogen clathrate which forms a structure involving 136 water molecules displaying a complicated geometry. The Canadian team examined the stability of the so called type II H$_2$/H$_2$O clathrate and compared with calculations. In their paper, the clathrate reaches the observed 1:2 H_2/H_2O composition at temperature $T = 250$ K, and pressure, P exceeding 1,000 bar.

Another exotic compound is the "metal-organic-framework compounds" MOFs, highly porous crystalline materials composed of metal clusters and organic linkers. These compounds are being studied by Omar Yaghi and his team at the University of Michigan, Ann Arbor. The Michigan group has identified up to 500 such compounds, one of which, $Cu_2(CO_2)_4$ units is based on and bonded by biphenyltetracarboxylic acid linker group and is reported to be able to contain up to 2 per cent wt. hydrogen. Many groups around the world are working on MOFs.

The general outlook for future work in the area of solid state hydrogen storage is related to better understanding of reaction mechanisms, improved kinetics and the study of reversibility in other complex hydrides.

An extensive list of complex hydrides and other hydrides which have been assembled by a remarkable American, Gary Sandrock and his colleagues, and can be found on the hydpark.ca.sandia.gov website.

The Grand Challenge of the modern day alchemists

Researchers are asking themselves if there perhaps is a solid state storage material that fits all the criteria of volumetric and mass percentages in addition to fast dynamics, but has not been "found", partly due to insufficient choices of the many combinatorial possibilities of, say, three element compounds.

Peter Edwards and his team at the University of Oxford, together with his research director Martin Jones and a partnership with Bill David and his team at the Rutherford Appleton Laboratory in Oxfordshire have started an exciting hunt for the holy grail.

Effectively, the team will sift through a "combinatorial haystack" of light-metal hydrides of varying combinations. This is done by applying vapour deposition onto different combinations of a number of elements using microfabricated hot-plate chips and by using the synchrotron facilities at the British facilities in order to examine thousands of combinations. When a given combination shows promising results, it is scaled up and tested in a larger bulk size. Johnson Matthey Company is partner and so is the Ilika company.

One of the interesting aspects of this literal "Grand challenge" is the use of automation making it possible to work a hundred to a thousand times faster than the manual alchemist could have done. The third-generation synchrotron sources such as the one in Oxfordshire are enormously more powerful than the old powder x-ray techniques traditionally used for such sample characterization. A sample which is running through hydrogen absorption or desorption process can be studied with a combined neutron diffraction and gravimetric analysis in-situ will quickly show if it holds a candidature for the Grand challenge. In summary, the use of the modern methods inherited from the silicon wafer technology and pharmaceutical industry are revolutionising the search for new compounds. It is to be expected that the compendium of ternary or binary phase diagrams of metals will need to be rewritten in the light of the discoveries of the Oxfordshire group [Zhitao et al. 2008.]..

Figure 36. The Rutherford Appleton Laboratory in Oxfordshire.

EFFICIENT HYDROGEN UTILISATION

Utilising the energy of the burning of hydrogen

When hydrogen and oxygen react by burning, about 33.6 kWh of thermal energy is released per kilogramme of hydrogen. This is usually termed "reaction enthalpy" and is different depending on whether the end product is water or steam as we discussed earlier. Considerable amount of hydrogen is burned in the chemical industry to produce heat. In the rocket engines of the space programmes of a number of countries, hydrogen is the fuel of choice. Hydrogen shows many advantages, for example regarding flame speed, i.e. the diffusion velocity of the flame in a hydrogen/air mixture. Hydrogen has a seven times faster flame speed than methane and about 3.5 times faster than town gas.

When pure hydrogen is burned in gas turbines one experiences relative absence of the problem of scaling or corrosion as would be in the case of fossil-fuels. On the other hand, the use of hydrogen may push the temperature limit to a range of high values where the tolerances of the turbine materials are challenged.

When hydrogen is considered for applications in classical drive trains of automobiles, it competes with fossil fuels in conventional combustion engines. Hydrogen can be used under partial load to power an Otto engine like in the petrol powered car, and shows similar efficiency as a Diesel engine. This has for example been utilised by BMW with impressive results.

Remember when we discussed the wide flammability of hydrogen (5 -75 per cent by volume). Therefore, operating an engine with hydrogen with very high excess air is possible. This is usually termed "a lean mixture of fuel and air". Petrol has a much narrower flammability range of only 2 – 8 per cent vol. Together with the advantage of the mixture range, hydrogen can be injected directly into the combustion chamber with the resulting engine performance improvements.

One autumn day in Cambridge, England, on November 27th 1820 the dons of the university assembled within the premises of the Cambridge Philosophical Society to hear a clergyman, Reverend W. Cecil, fellow of Magdalene College, give a long and detailed lecture about a rather unusual subject. The lecture was titled: "On the Application of Hydrogen Gas to Produce Moving Power in Machinery" and described an engine operated by the "Pressure of the Atmosphere upon a Vacuum Caused by Explosions of Hydrogen Gas and Atmospheric Air." The lecture, recorded in the transactions of the prestigious society, went on comparing coal based steam engines with the new concept which was believed to offer possibilities of much less cumbersome operation and speed.

This remarkable event in the history of hydrogen deserves to be accounted for whenever a brief account is given of the development of the subject. It counts as a unique event and can be viewed as a prophesy of the times we are trying to describe in these pages

Going back to our discussion of utilisation of hydrogen by burning and combustion, we have to note that burning hydrogen in combustion engines does not fully exploit the potential of the enormous energy involved. Tank to wheel efficiencies of hydrogen combustion engines suffer from the same problems as for petrol or diesel. Most of the energy released ends up in heat. Running a classical internal combustion engine on hydrogen yields somewhat lower efficiency than for an equivalent energy amount of petrol. However, best results of improving the efficiency of combustion engines running on hydrogen have been obtained in an European International project, HyICE, with participation of leading car manufacturers like BMW, Ford, Volvo and MAN. The target of this project has been to increase engine efficiency up to over 22 per cent which makes it comparable with the best diesel engines. Early 2007 this target seemed to have been reached successfully. Even more important, the NOx emissions of such engines, which always cause worries with hydrogen combustion, have been reduced quite dramatically. Let us, however, not forget that we have wasted about 4/5 of the energy into non-recoverable heat.

The classical burning of hydrogen does not utilise the full energy potential because of some fundamental drawbacks related to the very nature of combustion. So we may ask what other possibilities there exist for better utilisation of the energy involved.

In order to understand the scope and the possibilities we have to explore the findings of a remarkable scientist Josiah Willard Gibbs b.1839 who joined Yale University in his hometown, New Haven, Connecticut in 1871 and worked there until his death in 1903. In 1863, Gibbs was awarded the first Ph.D. degree in engineering in the USA from the Sheffield Scientific School at Yale. Five years after joining Yale, Gibbs wrote and published papers on the thermodynamics of "equilibrium of heterogeneous substances". In his astounding contribution to thermodynamics, Gibbs was able to include

Figure 37. J.Willard Gibbs.

aspects such as phase transitions, chemical energy and their effect on thermodynamic properties. Not only did Gibbs´ work include the effect of the pressure and volume aspects of a burning fuel, but also took into account the effects of the chemical energy at atomic level involved in the chemical reactions. The work of Gibbs in thermodynamics is sometimes compared to the work of Isaac Newton in mechanics. Gibbs' chief scientific papers appeared in the *Transactions* of the Connecticut Academy of Arts and Sciences which at that time had a relatively limited outreach in the world scientific community.

The contribution of Gibbs calls for a definition of, not only the enthalpy or heat energy of a substance, but also takes into account the contribution of entropy and temperature as we show in more detail in Cyber-Appendix VIII. For the time being we will define Gibbs free energy as the energy available to do external work, neglecting all work done by changes in pressure and/or volume. Another way to define it would be the maximum amount of chemical energy of the system that in a given situation can be converted into high-quality energy such as electricity.

So, within the realm of Gibbs´ theory and before going further, we need to consider a way to think of an electrochemical conversion of the chemical energy of a given fuel substance directly into electrical energy. The gateway of hydrogen as a fuel for electricity production had in fact been opened more than half a century before Gibbs by another remarkable scientist who now enters the scene with the resulting revolt in fuel utilisation theory.

The ultimate taming: The discovery and development of the fuel cell

The nineteenth century saw the birth of many of the electric devices so familiar to us today. Michael Faraday, a blacksmith´s son in England, showed how moving charges in magnetic fields produce a force; from which the electric motor was born - and Alexandro Volta introduced the voltaic battery.

A milestone in hydrogen development was set when a Walesman, William Robert Grove, born in Swansea in 1796, gave a lecture to the annual meeting of the Association for the Advancement of Science in Gibbs´ birth year of 1839. There he demonstrated a new concept of a gaseous "voltaic battery" which is commonly seen as the predecessor of fuel cells. Grove had graduated with a B.A. from Oxford seven years earlier and continued studying law at Lincoln´s Inn where he became Barrister in 1835. To touch on a light note, I sometimes wonder how much better the world would be if more lawyers were physicists!

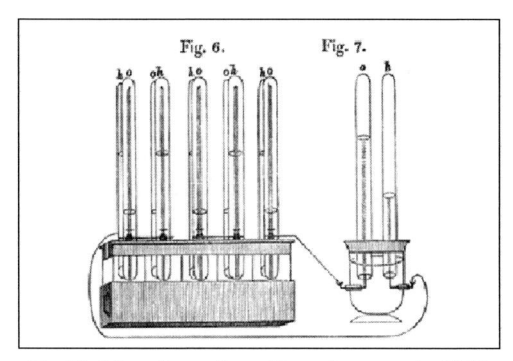

Fig.38 A key figure from Grove's paper in 1843.

At the time of Grove, electrolysis was well known from Nicholson´s original discovery in 1800 . When wires from the two poles of a battery are led into water to form electrodes, hydrogen and oxygen will develop at each electrode respectively. Grove had experimented with using platinum for electrolysis and found out that when steam came into contact with heated platinum, it decomposed into hydrogen and oxygen. He was using a row of glass tubes with the upper ends closed and the lower ends open, but immersed in diluted sulphuric acid as an electrolyte. When Grove connected together the leads from the two electrodes, he could observe a reverse electric current flow as if the hydrogen and oxygen stored in the glass tubes were returning back to produce water.

In those days these devices were only able to produce a faint electric current, but large enough to move a needle of a galvanometer. The set-up generated enough electricity that Grove could subsequently use the device to power a second electrolysis system.

Figure 38 appeared on page 272 of the *Philosophical Magazine and Journal of Science*, 1843, with William Grove's letter "On the Gas Voltaic Battery." Grove undertook the series of thirty experiments described in this letter when, as he depicted, "after my original publication I received a letter from Dr. Schönbein [Christian F. Schönbein (1799-1868)] ... [who] there expresses an opinion, that in the gas battery oxygen does not immediately contribute to the production of current, but that it is produced by the combination of hydrogen with water."

Grove spent a long time trying to perfect the invention and quickly realized that the bottleneck of his technology concerned the triple surface contact area connecting the gas, electrode and electrolyte. He later wrote that further work was needed in order to *"create a notable surface of action"*.

Grove´s invention had to wait another half a century before the next advances were made. Ludwig Mond and Charles Langer, who also worked in England, used the term *"fuel cell"* for Grove´s invention. They ran a number of experiments with the new device using hydrogen from a simple town gas system and atmospheric oxygen as a source. The main step they made concerned replacing the liquid electrolyte with a soaked up asbestos material named "plaster of Paris", the predecessor to, what we now call, solid electrolyte. The Mond-Langer device could produce some 1.5 Watts of electric power, but was deemed very expensive because of the requirement for platinum as electrode material. Most importantly though, the two gentlemen had certainly been able to create the aforementioned "surface of action".

Returning back to the work of Gibbs discussed in the previous section, we note that W. Ostwald made an observation of a crucial nature already in 1894 after examining Grove´s electrochemical process. Another scientist, C. Westphal, had manifested in 1880, that a very large part of the energy of the hydrogen/oxygen reaction in a fuel cell, could be converted into electrical energy. If the analysis is broken down further, it is possible to conclude that over 80 per cent of the reaction enthalpy can be converted into electricity. Electric motors are quite efficient and the comparison of fuel cell linked to an electric motor on one hand with a combustion engine on the other hand, is very much in favour of the fuel cell system. This defines a turning point in our discussion of the subject as a whole.

Moreover, this is also the transition point of utilisation from a conventional Carnot type of combusting hydrogen as a fuel to the modern idea of a Gibbs energy era of fuel cells for hydrogen energy conversion.

Somehow it became the destiny of British scientists to continue the remarkable work of the pioneers. In 1932 an engineer at Cambridge University, Francis T. Bacon, a descendant of the famous philosopher, tried to get around the problem of expensive platinum catalyst by building a fuel cell which was based on an alkaline electrolyte and made it possible to use inexpensive nickel as the catalyst. Bacon used about 200°C temperature and up to 40 atmospheres of gas pressure to show a considerable improvement of the performance of the fuel cell.

The World War II (1939-1945) ruled the scene and fuel cell development progress was delayed until about 1959 when Bacon demonstrated a stack of forty cells and was able to produce five kilowatts of electric power in a device with cells measuring about 25 centimetres in diameter. The cells were made of porous nickel and the electrolyte was made of potassium hydroxide. With the electric energy thus released Bacon managed to power a welding machine and a circular saw and to some extent a small fork lift. By all accounts, the Bacon fuel cell stack was already comparable to modern fuel cells. The main drawbacks were the demand for clean oxygen and hydrogen and the problem of CO/CO_2 fouling or poisoning, the terms used for the damaging effect of carbon oxides on the electrolyte. This was most often felt when the hydrogen utilised was made by reforming where remnants of carbon dioxide are left in the end product.

In 1959 the first fuel cell vehicle, a tractor, made by Allis-Chalmers farm equipment firm of Milwaukee, Wisconsin, was presented to journalists. The system never became commercial but certainly was a milestone on the way of utilising fuel cells.

The next step in the story of the development of the fuel cell was taken when General Electric of USA entered the fuel cell development in 1953. GE´s main interests were in the use of stationary fuel cells. A chemist, William Thomas Grubb working for GE, was in fact mostly interested in the rather far fetched concept of water softeners. In an attempt to "hook positive ions to polymer chains in a loose enough way that they can move around easily", Grubb proposed to make a fuel cell using sulfonated polystyrene resin as the electrolyte. In Grubb´s intuitive picture, hydrogen ions would migrate through the membrane and combine with oxygen on the other end. This first *PEM, Proton Exchange Membrane, cell* was completed in 1954. Cationic ion exchange compounds, of which sulfonated polystyrene is a classic example, are sold in high-volume markets for water purification devices. Essentially, such a system, when charged with protons, exchanges protons by removing other cations, say Ca^{++}. The GE work built these materials into sheet form.

Let us at this point make a break and consider very carefully the implications of the fuel cell discovery. It was known for years that certain ions can migrate through crystals. An example is the transport of the silver ion through solid silver iodide, AgI. It is easy to show that one can transport silver between two silver sheets sandwiched on both sides of a solid AgI wafer. However, this process requires temperature high enough (600°C) to provide translational energy to the much smaller silver atom within the AgI crystal structure. Transport is driven by an electric current. Without realizing it, these scientists had for the first time in history been able to transport a proton flux through a solid chemical polymer. These ion exchange membranes are physically similar to the case of AgI mentioned above. The sulfonic acid groups, $-SO_3^-$, are chemically bound to the polymer back bone and are thus immobilised. However, protons can move from one sulfonate group to the adjacent one, resulting in a flux of protons. Nevertheless, the rate of that flux is always slow contrasted to the far lighter electron which moves through wires again, probably using a "hopping" sort of mechanism.

The early Grove experiments were made with aqueous sulfuric acid solutions. Indeed that worked, but not well. The challenge really is to minimize the distance that protons must travel between the two electrodes. The goal has always been a "zero gap" design, that has the anode electrocatalytic layer and the cathode electrocatalytic layer in close proximity, so that the resistance losses, the result of slow proton transport, are minimised. However, one must avoid mixing the fuel with air, because there is another certain result, combustion, when that occurs. (The necessary fuel cell catalysts are all excellent combustion catalysts.) Consequently there needs to be a gas barrier that precludes reactant mixing. So, this fuel cell separator needs to provide both ionic transport and reactant separation. Obviously this involves an engineering tradeoff, because an optimum transport device will be very thin, and an optimum gas barrier will be less so.

The reaction mechanism is not simple. The hydrogen anode reaction involves first the chemical adsorption of the H_2 molecule onto a catalyst surface. The molecule splits, into two adsorbed hydrogen atoms. An electron is removed, which flows through an external electron conducting pathway, leaving a bare proton. The proton reacts with a water molecule forming a hydrated species, mostly H_3O^+. That "hydronium ion" migrates through the proton exchange membrane, driven by a charge gradient. Oxygen also adsorbs on electrocatalysts, this time on the surface of the cathode catalytic material. That reaction of oxygen to form water is complicated and can occur through a network of parallel reactions. In the ideal case, a di-oxygen molecule reacts to add four protons and four electrons (accepted from the external electronic flow), to form two water molecules. This model illustrates one complexity. Water is a product at the air electrode, but is a reactant at the fuel electrode. Fuel cell operation does involve water transport processes. Although the picture is that the catalyst remains useful, certainly the electrocatalysts are necessarily chemically active. The anode catalyst behaves in a parallel manner to metals that form chemical hydrides, while the cathode catalyst, when it "adsorbs" oxygen, is clearly involved with chemically binding to oxygen. So, the correct conclusion is that another series of reactions are necessarily involved which regenerate catalyst surfaces, so that a casual observer would conclude that the metals are not involved with the process.

History finds a multitude of unique ways to fold out. When the two superpowers of the 20th century started into race for the space, fuel cells again became a matter of great interest as the need grew for compact energy conversion devices onboard spacecraft. The USA space program tended to favour fuel cell technology, even though the USSR activities utilised batteries, although fuel cells were thoroughly explored by both groups. A General Electric team led by Leonard Niedrach started simplifying the manufacturing technology by using platinum deposited metal mesh directly bonded to the polymer membrane.

The very early US Gemini programme used a PEM system for the early flights, based on sulfonated polystyrene hardware. This worked well, although today the Gemini fuel cell performance looks embarrassing low, with a very large internal cell resistance. There was good reason to look for another fuel cell design. The Apollo programme used an alkaline fuel cell system, a low pressure version derived from the work of Bacon. General Electric, still pursuing PEM designs, sought the assistance of a chemical company, the DuPont Corporation where ion exchange membranes were used, trademarked as "Nafion", which had been developed for a different application. Nafion is based on polyperfluorohydrocarbons, much like the Teflon "frying pan" polymer. General Electric, using the Nafion materials, trademarked their hardware as "SPE", a solid polymer electrolyte fuel cell. Fifty years later, NASA still flies with KOH fuel cells, the "orbiter power plant", while PEM fuel cells have been the focus of different applications.

In 1983 a team headed by Geoffrey Ballard of Canada received a Canadian government´s request to develop a PEM fuel cell. The impressive company of Ballard in Canada currently is a leading company in the fuel cell business. GE transferred its PEM technology which in the end became a part of the UTC, another impressive company, the United Technologies Corporation. Many notable companies can be mentioned as producers of fuel cells. We will come to them in the more specialized discussion about types of fuel cells.

The development of other types of fuel cells started in the late 1930s with the work of Emil Bauer and H. Preis in Switzerland on solid oxide electrolytes in fuel cells and O. Davtyan of Russia in 1945 on mixing carbonate and oxides with a sand separator which formed a work basis for the post-war fuel cell work. In the next pages we will look deeper and more systematically into different types of fuel cells.

The fuel cell menu

We can summarise that a fuel cell is a static electrochemical device (no moving parts) that converts the chemical energy of a fuel, such as hydrogen, and an oxidant, such as oxygen, directly to electricity and heat. The principal components of a fuel cell are catalytically activated electrodes for the fuel (anode) and the oxidant (cathode) and an electrolyte to conduct ions between the two electrodes.

Like a battery, the fuel cell can convert chemical energy into electrical energy. However, instead of using electricity to recharge a battery, a fuel cell requires continuous input of fuel and oxidant to operate.

A fuel cell will compete with other types of energy conversion devices, including gas turbines, internal combustion engines in automobiles, and batteries for, to name an example, laptops.

A basic constituent of a fuel cell are two electrodes, the anode and the cathode. The electrolyte is situated in between the electrodes and has the role of transferring ions in either direction while an electric current balances the ionic flow. The electric current can be applied to drive an external circuit such as an electric motor. .

Many fuels are possible for powering fuel cells. The most notable is molecular hydrogen. Hydrazine, N_2H_4 is also a possible high yield fuel which was extremely popular during the latter third of the 20th century but suffered from toxicity problems. Natural gas and petroleum are potential fuels, even in their natural state, but may also need reforming to make hydrogen as we have seen earlier.

We will distinguish between two main types of fuel cells. Other less common fuel cells will be studied separately. On one hand there are the **PEM and phosphoric acid fuel cells**, where protons move through the electrolyte to the cathode, producing water and heat. Secondly, we have **alkaline, molten carbonate and solid oxide fuel cells,** where negative ions travel through the electrolyte to the anode where they combine with hydrogen to generate water and electrons.

The electrons from the anode side of the cell cannot pass through the electrolyte to the positively charged cathode: they must travel around via an electrical circuit to reach the other side of the cell. This is in fact the electrical energy output of a fuel cell and can be transferred by an electron passing through a load, an electric motor or the like.

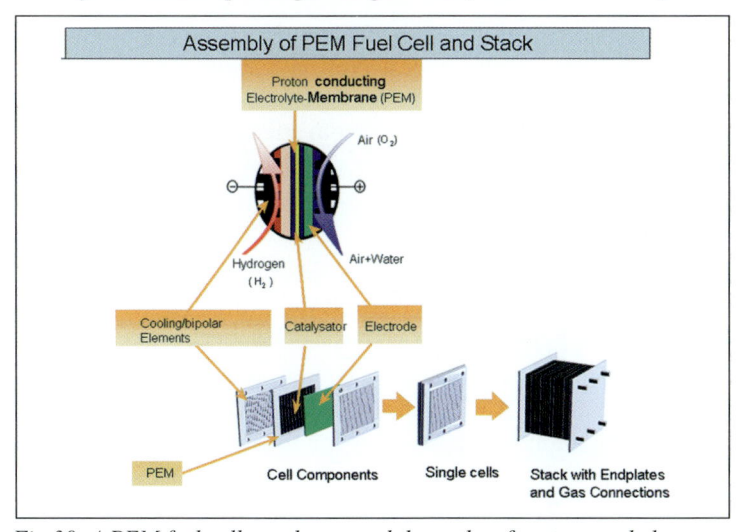

Fig.39. A PEM fuel cell membrane and the paths of protons and electrons.

The main features of a fuel cell are that it contains no moving parts; it therefore has long lifetime reliability, at least in theory; and it has a high efficiency; with no Carnot limitations it is able to reach efficiencies in the range 40-70 per cent. Last but not least, can have low emissions as we will examine in the cases below. The highest efficiencies of fuel cells are attained at low temperatures. The competitors, heat engines, usually have lower efficiencies, and they follow Carnot´s law. Figure 40 shows a comparison between the efficiency of a Carnot heat engine and an ideal fuel cell. The fuel cell starts with very high efficiency at low temperatures and drops off slowly as the temperature is increased. On the other hand, the heat engine starts very low and gradually increases its efficiency until about 1000°C where the two become comparable.

The usual method in classification of fuel cells is to use the *fuel* itself as a basis as well as the type of *electrolyte*.

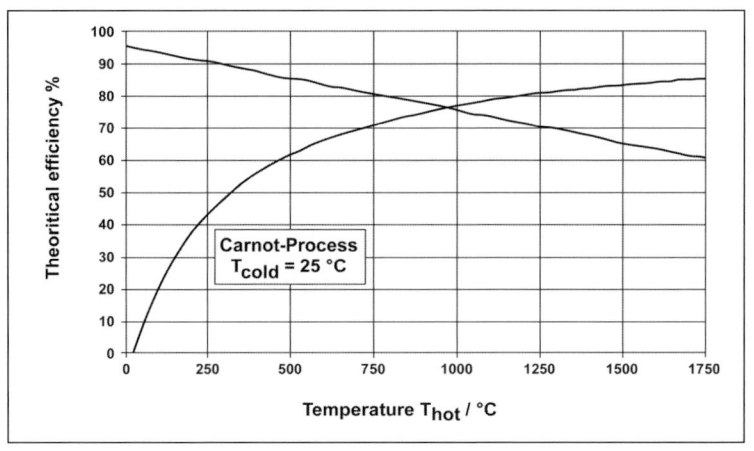

Fig. 40. Comparison between the efficiency of a Carnot heat engine and an ideal fuel cell as a function of temperature.

If we base our classification on the fuels, we can have *direct* fuel cells where hydrogen is fed directly to the anode; or we can have *indirect* fuel cells where external reformers are used to supply the hydrogen to the anode. Finally we can have the *regenerative* type where the fuel cell product is reconverted into reactants and recycled.

In the second classification all is based on the *electrolyte*. We then have polymer electrolyte (PEFC) such as *Proton Exchange Membrane* PEMFC operating at 80°C with the proton as the main moving "ion".

As we move up in temperature and to phosphoric acid, which we remember from the original Grove experiments, we come to PAFC with an operating temperature of about 200°C and proton being transported through the electrolyte.

Moving to fuel cells based on negative ion transport we come to alkaline electrolyte and we get the *Alkaline* AFC, with operational temperatures around 80-100° C. Here the negative OH^- ion is transported through the electrolyte.

Molten Carbonate MCFC operate at 650°C with a carbonate ion on the move and finally *Solid oxide* fuel cells SOFC are based on solid oxides, oxygen ions on the move, and operate in temperature ranges from 800-1000°C.

Planet Hydrogen

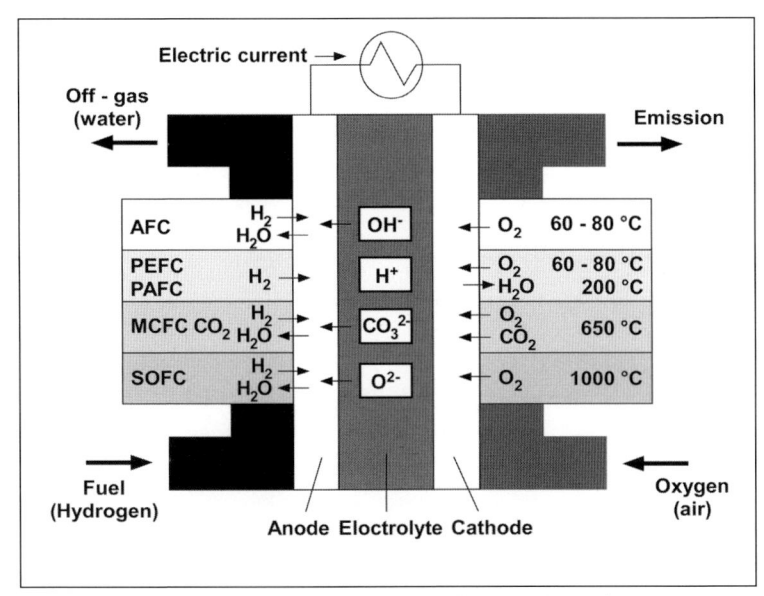

Fig. 41. Different types of fuel cells viewed together.

Figure 41 shows all the different types of fuel cells in the same schematic figure. We could also divide the fypes of cells into stationary versus mobile or portable. The stationary fuel cells utilise the fuel and yield both electric energy and heat. The mobile ones are intended for transport, while the portable ones destined for electronic or electric appliances. Their engineering challenge is how to capture and make use of excess heat.

Proton Exchange Membrane Fuel Cells

We will start with the PEMFC which in many ways represents the full taming of the proton. Figure 39 and 42 show the PEMFC the Proton Exchange Membrane fuel cell. The electrolyte is made of a material permeable to protons. This material could for example be made of Nafion adjacent to carbon which has been coated with platinum catalyst. At the anode, which has etched channels to disperse the gas, the hydrogen molecule is decomposed into two protons and two electrons. At the cathode oxygen is reduced. Also here, etched channels distribute oxygen to the surface of catalyst. The catalyst is rough and porous to increase contact area; often platinum powder thinly coated on carbon paper. This is focus on increased contact area is effectively meeting William Grove´s original remark and thought about "a notable surface of action".

The PEM fuel cell has quick startup and is the primary candidate for the automobile industry. It runs on pure hydrogen and is only tolerant to ppm levels of CO_2. The exhaust is pure water. The steam coming of the typically 80°C hot PEMFC is quite saturated and visible. Currently development work is concentrating on Pt/Ru catalysts that are more resistant to CO.

One of the most notable and successful uses of PEMFC in large demonstration projects has been in the CUTE/ECTOS project within the European Union and the Step

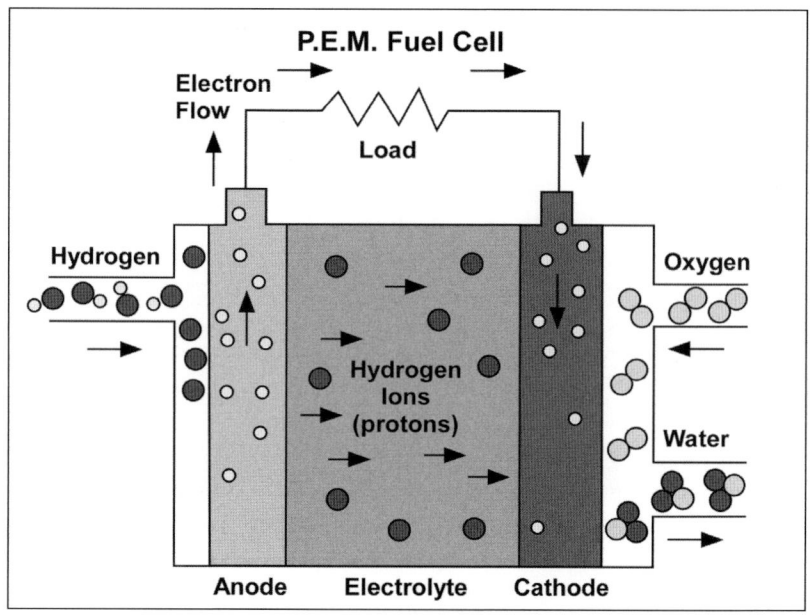

Fig. 42. Principles of a PEM, proton exchange membrane fuel cell.

project in Australia. 30 hydrogen-fueled Citaro busses powered by a Ballard PEM fuel cell system have been used in this project in 10 cities since 2003. A part of the project was extended in 2006 and will be described in the relevant sections of this book as well as other bus programs underway in Japan and the United States.

The emphasis on vehicular development involves most major global automobile corporations. Many of these are part manufacturers and suppliers. Historically, about 80 per cent of fuel cell system failures were the result of balance-of-plant component failure, not fuel cell failures. There has been enormous progress in engineering of high-quality fuel cell system parts, both for the vehicular sector, and other fuel cell application sectors.

A discussion of fuel cells and applications

A fuel cell vehicle is, of course, an electric vehicle. Indeed, almost all fuel cell vehicles are now designed as fuel cell-battery hybrids. PEM fuel cell engineering has led to advances resulting in increased performance and power density. Volumetric power density exceeding 2 kW/l have been demonstrated. A variety of new membrane materials have also been produced which have important advantages over the earlier Nafion types. Importantly, the new membranes are far less permeable to oxygen, so much thinner sheets are useful. This advance resulted in high voltages, even at high current density operation.

Because of these advances a variety of other applications are being pursued. Electric trains are under development in Japan, where fuel cell (hydrogen-powered) rail cars are being designed by two railway firms. These rail cars are also fuel cell battery hybrids, where the energy discharged during deceleration is used to charge the battery for acceleration. Importantly, such hardware eliminates the need for electrical supply systems, a significant part of the capital cost of electric railway systems. Work on fuel cell powered boats is also underway, and recently a German consortium has flown a prototypical small aircraft, perhaps a prototype of a future surveillance platform.

Another important PEM fuel cell application is in lift trucks. These are frequently used and operate for many hours each day. These systems utilise pressured hydrogen storage tanks, and have large advantage in clean and quiet operation, and in rapid refueling contrasted to battery driven trucks.

However, PEM fuel cells are not limited to vehicular applications. A large project in Japan deploys thousands of small, distributed residential systems. These are in the main reformer-based PEM systems. About 50 per cent are fueled using natural gas, and the rest fueled using a light diesel fuel (kerosene), or propane-butane. These devices could be considered as water heaters which make electricity as a byproduct. The fuel cell system is not particularly efficient. A 1-kW fuel cell electrical output results in perhaps a 3-kW thermal output. Some of this heat emanates from the fuel cell, some from the fuel processing hardware. These Japanese-designed and built systems consist of two small cabinets. One cabinet houses the hot water storage tank, and the second contains the reformer-fuel cell hardware. These devices are, in the main, grid connected. The most interesting operational mode is to follow the thermal load, usually water in a hot water tank. When the tank cools to a certain level, the fuel cell system operates. Electricity is inserted on the grid, and the product heat transferred to the water tank. Hot water is used for space heating and for domestic hot water purposes. The combined efficiency of these devices, electricity plus heat, is in the 80 per cent region. Importantly, the fuel cell operates only when both heat and electricity are generated, and therefore fuel is used efficiently. In this way, much less fuel is used than with usual (separate) hardware. This programme, which is now in full operation, has been successful. Interestingly, thousands of these small hydrogen generators (all steam reforming) have been deployed in highly populated residential areas, a very informative demonstration that hydrogen can be successfully produced in a very distributed manner.

PEM fuel cells are also being used successfully as replacement for batteries used in portable power applications. The majority of these commercial systems utilise pressured hydrogen, although several methanol devices are also being evaluated and marketed. Some of the methanol systems are "direct fueled" and some involve a miniature methanol reformer which generates a hydrogen-carbon dioxide fuel. Developed products are small enough to replace laptop computer batteries. Interestingly, one significant application for these fuel cell devices is as battery chargers. Thus, although some saw fuel cells as competitors with existing battery hardware, the result is that fuel cell systems are making existing rechargeable battery designs more useful.

Although PEM FC technology continues to make large strides towards significant market penetration, technical problems still persist. Capital costs remain high, and durability at times is not sufficient. Even so, many technical problems have been overcome already, and the few required improvements could happen quickly. The emphasis on cost reduction is certainly warranted, however it must be remembered that for some applications, such as city buses, life-time costs typically are dominated more by maintenance costs than so called "first costs". The early testing data suggest that fuel cell buses actually could decrease maintenance costs, probably compensating for any additional "first cost." Moreover, the emphasis on exceedingly-long life time for fuel cell systems may be unrealistic. A fuel cell bus, for instance, has a number of parts that are routinely changed as part of the scheduled maintenance. Many parts are changed repeatedly over the useful bus lifetime. A fuel cell stack is just one of the many components which may need to be changed following a thoughtful maintenance programme—fuel cells, unlike all other electrochemical devices, should not be expected to last forever.

Phosphoric Acid Fuel Cells

Figure 43 shows the *Phosphoric Acid Fuel Cell, PAFC*. It contains highly concentrated H_3PO_4 in Teflon-bonded silicon carbide SiC matrix. More modern versions utilise phosphoric acid adsorbed into a porous polymeric sheet made of polybenzimidazole. There are hundreds of such fuel cells in use worldwide and they are often called the first generation of modern fuel cells. They are typically used for stationary power generation and a few PAFCs are used to power large vehicles such as city buses. PAFC are much more tolerant to carbon monoxide impurities in the anode feed that have been reformed into hydrogen compared to PEM fuel cells. When used in combined heat and power applications PAFCs have shown up to 85 per cent efficiency. Their efficiency in electric production alone is about 40 per cent.

These systems, deployed during the 1980s, experienced technical problems. Indeed, PAFC hardware was incorporated into multi MW demonstrations. Bipolar plates

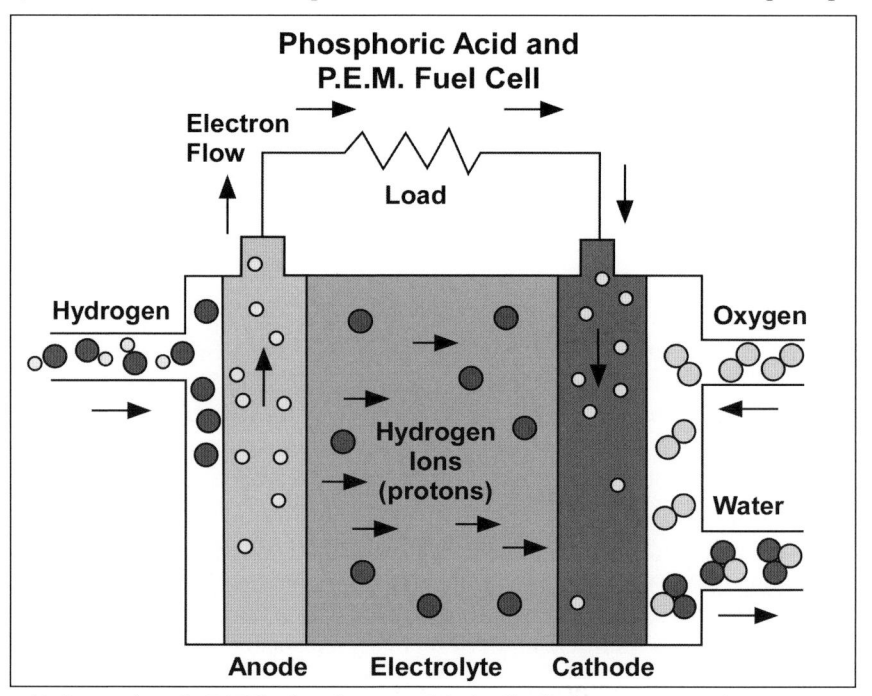

Fig. 43. Principles of a PAFC, phosphoric acid fuel cell which in essence is the same as PEM but the electrolyte contains highly concentrated phosphoric acid.

Planet Hydrogen

were made of graphite, and corrosion on the anode side was a life-time concern. However, more recently, as a result of greatly improved fluid dynamics and system control, life time and performance have shown good progress, and PAFC is again being marketed by Japanese companies.

Alkaline Fuel Cells

Figure 44 shows the principles of the *alkaline fuel* cell AFC. It contains concentrated potassium hydroxide, KOH (35-85 wt. per cent) in asbestos matrix and can use variety of non-precious metals as catalyst at anode and cathode. This fuel cell has been used in the US space programmes for 45 years. A high-temperature version operating at temperatures from 100°C to 250°C. Then there is a newer version able to operate at lower temperatures from room temperature up to 70°C. In space applications, the AFC fuel cells have demonstrated efficiencies up to 60 per cent. It has long been understood, that a variety of transition metals are useful cathode catalysts, and this "non-platinum" feature is a significant advantage.

Current space designs utilise a hydrogen recirculation stream—hydrogen is pumped continuously through a loop which includes the fuel cell— both for water removal and thermal management. Liquid water is a challenge in gravity-less space, and the design includes features which accomplish water management. The KOH solution is far less corrosive than a similar acidic solution and common metals, such as nickel can be useful for cell parts. Inherently, the oxygen reduction in alkaline media occurs with less

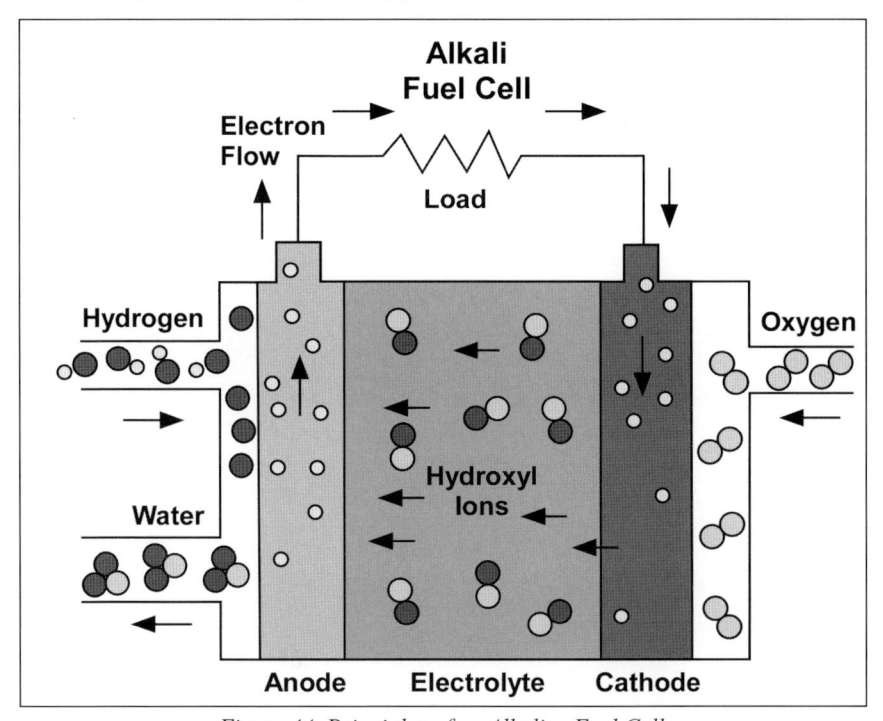

Figure 44. Principles of an Alkaline Fuel Cell.

voltage loss than found in acidic solutions. On the other hand, the resistance of the electrolyte in KOH hardware, results in more voltage loss at higher current density levels than found in modern PEM hardware. Among the disadvantages are sensitivity to CO_2. Carbon dioxide at 400 ppm can be easily removed for the cathode feed. However, the use of a hydrogen-carbon dioxide mixture (product of a fuel processor) can only be done if the electrolyte is continuously regenerated. Even so, for hydrogen systems, AFCs are technically feasible. One challenge, of course, is that product water dissolves in the KOH electrolyte. That water needs to be removed essentially at the same rate as it is generated.

The space hardware operates at a temperature-pressure location, where water is automatically removed maintaining the electrolyte concentration. That is one approach. There are other ways to accomplish this step. The undesired consequence of not removing water is the creation of a large volume of dilute potassium hydroxide!

Molten Carbonate Fuel Cells

Figure 45 shows the principles of a *Molten Carbonate Fuel Cell* MCFC developed for natural gas and coal-based power plants for utility, industrial and military applications. The electrolyte is made of molten carbonate salt mixture suspended in porous, chemically inert ceramic $LiAlO_2$ matrix.

The MCFCs work at quite high temperatures around 650°C with non-precious metals as anode and cathode catalysts and thus reducing costs. The beauty of the MCFC is that it does not require external reformer to convert the fuel to hydrogen. Due to high operating temperatures, the fuels are converted to hydrogen by internal reforming which surely reduces costs. Fuels can be H_2, CO, natural gas, propane or diesel.

The reported efficiencies are near 60 per cent, appreciably higher than the corresponding ones for PAFCs. In some cases, waste heat is recovered yielding fuel efficiencies up to 85 per cent.

The world´s first fuel cell which generates electricity and heat from biogas at an industrial scale was installed in the county of Böblingen in Germany in 2006. It was designed and built by the company MTU CFC Solutions. It is of the molten carbonate type and is linked to an aerobic household waste fermentation plant and produces some 250 kW of electricity, 120 kW of heat at an operating temperature of 650°C.

When the fuel cell is supplied with the mixture of methane and CO_2, the hydrogen of the methane reacts at the anode with the carbonate ions of the electrolyte to form water and carbon dioxide. Thereby, electrons are released. Together with the atmospheric oxygen, the carbon dioxide is fed to the cathode. By using electrodes, new carbonate ions are constantly produced at the cathode where, in turn, heat is released. The ions travel through the electrolyte to the anode – this is the basis of electric power production within the device.

Electric efficiency of this plant is approximately 47 per cent and thermal efficiency 23 per cent, resulting in a maximum total efficiency of about 70 per cent. A great opportunity exists for the use of MCFCs with industrial gases and for example sewage gas.

Planet Hydrogen

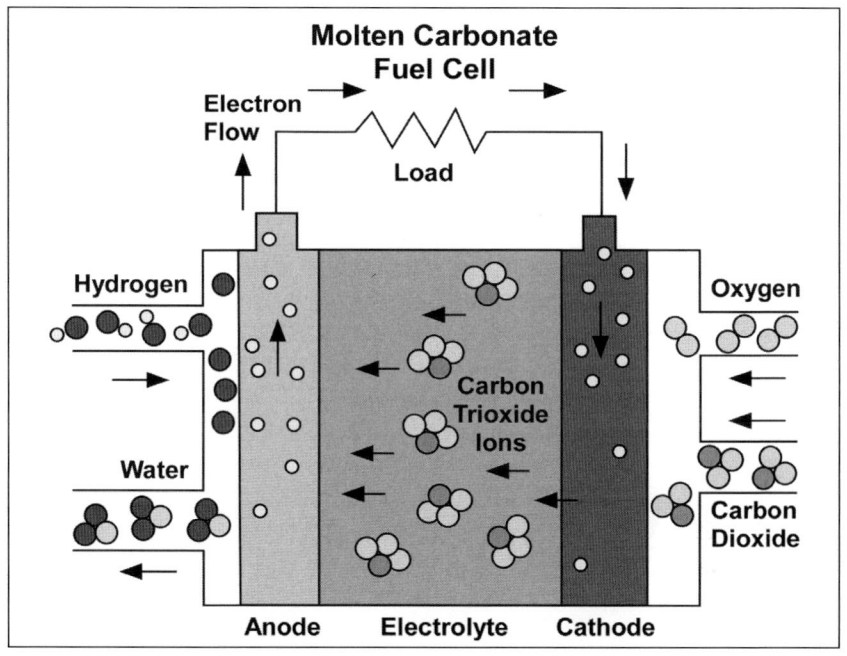

Fig. 45. Principles of a MCFC, molten carbonate fuel cell.

Solid Oxide Fuel Cells

Solid oxide fuel cells utilise a ceramic solid material as the electrolyte. These solid state electrochemical devices are already widely used as sensors—for instance an automotive exhaust gas oxygen sensor is made utilizing the same type of material. Voltage is read with one side of the hot ceramic exposed to exhaust gas and the other side ambient air. Output voltage is a measure of exhaust oxygen content. The fuel cell works in a reverse manner.

Figure 46 shows the principle of a Solid Oxide Fuel Cell SOFC. An SOFC uses a solid, non-porous ceramic compound as the electrolyte, a Y_2O_3 stabilized zirconia ZrO_2. The anode catalyst materials are $Co-ZrO_2$ or $Ni-ZrO_2$ whereas the cathode is made of Sr-doped $LaMnO_3$. The catalysts are also ceramic materials, and the device is formed as a three layer ceramic structure.

Since the electrolyte is solid, the cells can be constructed in a variety of ways. They are usually of two general types: tubular or compressed plates. Their operating temperature are highest among fuel cells, up to 1000°C with efficiencies around 50-60 per cent. By capturing the system waste heat for cogeneration, the overall fuel cell efficiency projects to 80-85 per cent.

Like in the case of MCFC, the high temperature tolerant ceramic removes the need for a precious metal catalyst which again reduces cost. The high temperature also allows the SOFC to operate directly on fuels which in turn are reformed internally. Another feature is that the tolerance of SOFC to sulfur is the highest among all fuel

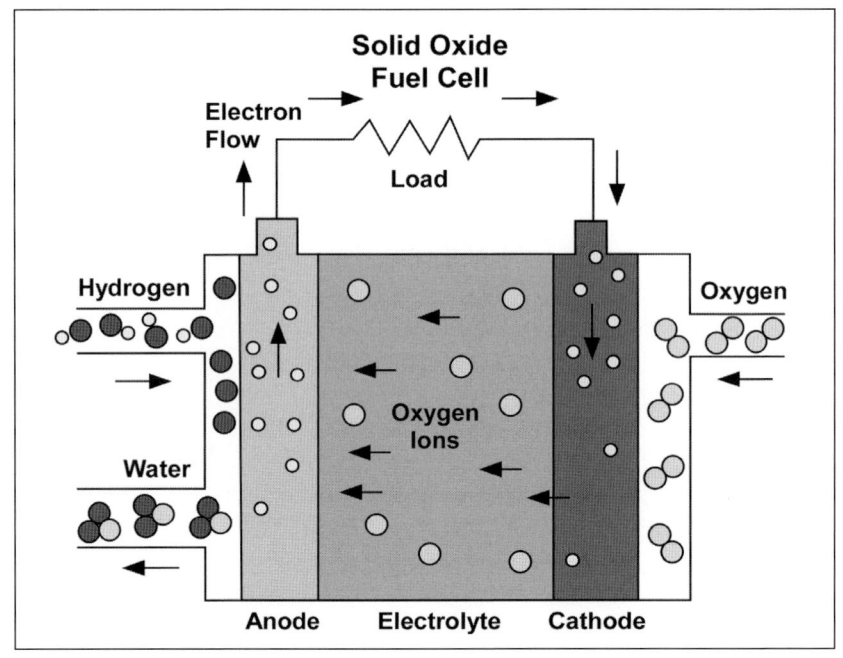

Figure 46. The principles of a Solid Oxide Fuel Cell.

cells. Natural sulfur content of fossil fuels can often be tolerated. Another beauty is their tolerance to carbon monoxide which also can be used as a fuel! This allows SOFCs to be used with gases made from coal. However, SOFC devices do not consume sulfur—the sulfur entering the process tends to come out.

As an example of a SOFC Combined Heat and Power system (CHP) we can take a 100 kW SOFC system demonstrated by EDB/Elsam in Holland. Figure 47 shows the system that is composed of a SOFC fuel cell connected to a small heating system intended for space heating and hot water supplies of a number of dwellings. The main fuel is natural gas. Outside air is heated with the use of a heat exchanger which in turn receives hot gas from the fuel cell. In case of start up of the fuel cell it is possible to preheat the air. Start up times can vary. A direct current of about 120 kW power is produced. Current inverter converts the direct current to alternating current at about 400 V and 109 kW.

The electric efficiency of this system is about 46 per cent. Furthermore, the heating provided by the unit makes the total efficiency of CHP of 76 per cent. One of the successes of this demonstration is a very low NOx and CO content of the effluent gas as well as having been working well over three years.

SOFC technology has had a long and complicated development path. Although the devices appear simple, that appearance is deceptive. As with all fuel cells, one must provide both pathways for electrical conductivity and ionic conductivity. It is quite possible to make conducting ceramics, which are indeed required for catalyst supports. These are typically mixed valence state metal oxides. However, it is also possible to oxidize these materials, and thus degrade electronic conductivity. The coefficient of

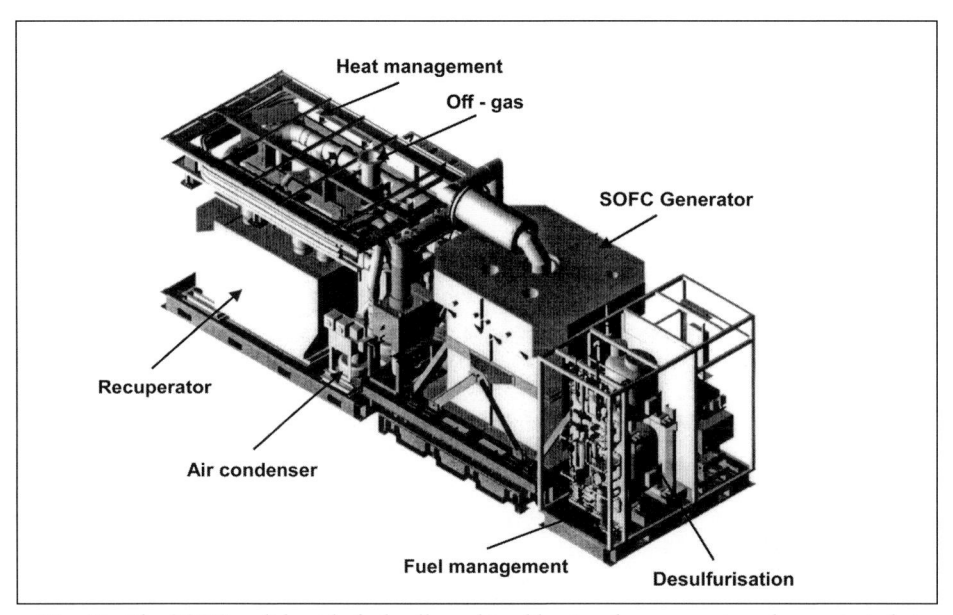

Fig. 47. The SOFC, solid oxide fuel cell combined heat and power system demonstrated at EDB/Elsam in Holland.

thermal expansion is also a concern. These layer structures will expand when heated. Even if the layers have well-matched coefficients of thermal expansion, it is possible that during heating one section of a single part will be hotter (and larger) than another. Sealing is also a continuing concern. The electrolyte must be impermeable and gases also must not mix at the electrolyte edge. In short, SOFC technology involves a series of very challenging engineering tasks. These concerns have been understood for decades, but it has only been a short time that significant progress has been achieved.

The most promising designs for commercial SOFC are tubular. However, round tubes are maybe not the best. Several developers are extruding flat tubes (ratio of width to height of 10:1), rather like planer hardware, but with the edges thoroughly sealed. Such tubes typically have the air electrodes on the outside, while the fuel (anode) electrode is in the interior of the tube. Significantly, the device is mounted only on one end, so that the tube is free to expand, so startup can be rapid. Major companies with decades of excellence in ceramic engineering are now involved. Significantly, these new designs use electrolyte formulations that show adequate ionic conductivity at 750°C, totally changing the degradation rates caused by high temperature ceramic oxidation and by thermal migration of metallic components. Several of these new designs are in the marketplace, and SOFC technology is now receiving a very large and rapid infusion of investments for commercialization,

Direct Methanol Fuel Cell

The final fuel cell type we will discuss is the *Direct Methanol Fuel Cell*. In Cyber-Appendix III we discuss the working of a reformer, a device for turning hydrocarbon fuels into hydrogen.

Direct Methanol Fuel Cells DMFCs are powered by pure methanol mixed with steam and fed directly into fuel cell anode. Figure 48 shows the principles of its operation. On the anode side, methanol and water are mixed and fed to the cell. On the cathode side ambient air circulates. The oxygen reacts with arriving protons and electrons to form water. A number of materials can be used for electrolyte, e.g. Nafion as in PEM fuel cells. The water management of the methanol and water mixture presents a problem for developers and has led to a variety of experimentation with different materials.

The elegance of DMFCs is that they can be used directly with liquid methanol storage and are, for example, considered very important for small appliances, laptops and the like. They suffer from slow reaction times which manifests itself with low voltages and low efficiencies, sometimes around 25 per cent. Compared to the best of batteries, for example lithium-ion batteries, the DMFC can yield up to eight times more energy from a given mass of storage – in this case battery or methanol tank respectively.

Some debate has been governing the acceptance of DMFCs by industry because of an alleged toxicity of the methanol as a fuel. Many believe that DMFCs will consti-

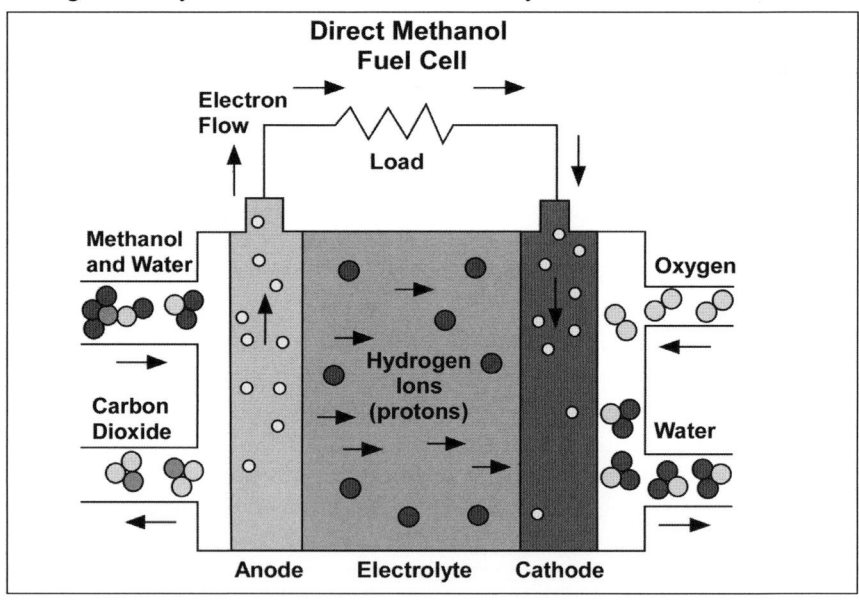

Fig. 48. Principles of a DMFC, direct methanol fuel cell.

tute the first large scale commercial uses of fuel cells. Applications already in use include laptop computers, scooters and small remote power systems especially uses of military nature.

Cyber-Appendix VII describes in more detail some fuel cell fundamentals such as fuel cell performances etc. The challenges facing this technology concern mostly costs which also have to do with the use of expensive materials. The size and weight also have to come down to meet the requirement by automobile producers. Thermal and water management also possesses a challenge. In the past five years a lot of progress has been made in this area and the guaranteed frost tolerance of automotive fuel cells is going 20°C or more below freezing as a result of proper water management.

Regarding commercialization of fuel cells, the aim of the industry concerning fuel cell lifetime is more than three years for automobile applications and exceeding 10 years for stationary applications. As regards performance, aims are to obtain more than 60 per cent efficiency for the automobile and power to mass ratio of more than a kW per kilogram. Start-up times are required to be less than one minute.

For commercialization of stationary fuel cells, acceptable price target is in the range $400-$750 /kW, still a bit on the high side. Lifetimes have to improve to about 50.000 hours of operation before they become finally competitive.

Ballard has achieved impressive results with for example reducing the need for platinum used in their fuel cells. In 1994 the use of Pt amounted to about 8-10 mg/cm^2; In 2004 it had been brought down to about 1 mg/cm2; In 2006 it was about 0.3-0.5 mg/cm^2 and heading towards less than that value after 2010.

Ballard has reported that it is aiming at a price target of **$120/kW** for its automotive fuel cell systems when volume production starts in 2008. Let us keep in mind that the cost of combustion engine power plants is about $25-$35/kW. The lifetime target set by Ballard is more than **6,000 hours**, equivalent to about 300,000 km. So far as fuel infrastructure was concerned, Ballard has said that from 2008 it expected fuel cell vehicles running off hydrogen would be tested in expanded fleets.

Finally a short note on fuel cell systems. A fuel cell drivetrain or system requires a number of components in addition to the fuel tank, the fuel cell and the electric motor. Both fuel and air need to be moved around the fuel cell. To perform this, various types of pumps, blowers and compressors have to be used. Figure 49 shows the process flow

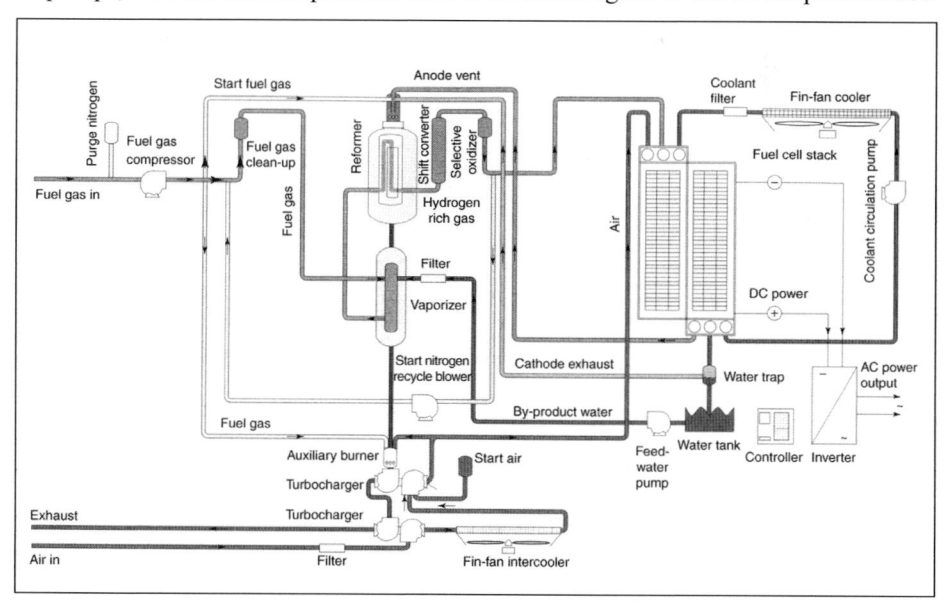

Fig. 49. Process flow diagram for a Ballard 250kW PEM fuel cell plant for automobile applications of the type used in the European fuel cell bus projects.

diagram for a Ballard 250 kW PEM fuel cell plant similar to that used in the CUTE/ECTOS buses.

In many systems, turbines need to be used for harness the energy of the exhaust gas. Then ejectors may be needed to circulate the hydrogen if it comes from a high pressure store as well as recycling anode gas. Various types of cooling fans and blowers are needed. In smaller systems there is often a need for membranes or diaphragm pumps to pump reactant air or hydrogen. The reader is referred to an excellent chapter by J. Larminie and Andrew Dicks (2002) on this subject. Also, the treatment of the delivery of fuel cell power as regards DC regulation and Voltage conversion, inversion and electric motors involving fuel cell/battery hybrid systems are discussed in depth. We address some of the fuel cell fundamentals in Cyber-Appendix VII.

PART IV
HYDROGEN INFRASTRUCTURE AND SOCIETY

Hydrogen entering society

Up towards the end of the 20th Century hydrogen energy had not gained any political weight and was merely pushed by technological considerations and motivated by scientific excellence. With the advent of the concept of a hydrogen energy economy, socioeconomic issues gained importance and today no serious hydrogen demonstration project does not include socioeconomic dimensions.

As the technology gained maturity it became necessary to deliver innovative solutions and products that could contribute to a successful market introduction of hydrogen and fuel cells into an emerging new market. Socioeconomic aspects concern, for example, the policy framework that surrounds a given technology, including policy targets and the resulting legislation.

Governments can in principle address hydrogen energy from three different angles: with soft policy, including declarations, target dates etc; with hard policy, with direct r&d funding, public-private partnerships, codes and standards and so forth; and finally with indirect policy, involving full-cost energy pricing, environmental regulations, tax incentives, public education and the links to climate change policy and energy security.

The branches of socioeconomic considerations regarding hydrogen are very far reaching because they involve the science system, societal demand, socio-cultural preferences as well as our societal values. Currently, care for the environment has become a very important element in the marketing matrix for hydrogen.

The technological transitions foreseen with the advent of hydrogen will most likely be unique but could be compared with any long-term and large-scale historical changes in the sociotechnical systems. In a beautiful analysis of technological transitions Frank Geels of the University of Twente, Holland, has used three interesting historical transitions to illustrate the conceptual perspective. Geels addressed the transition in oceanic shipping from sailing ships to steamships. Furthermore, he studied the transition in urban land transportation from horse-and-carriage to automobiles and finally he narrowed his focus down to the transition in aviation from propeller-piston engine aircraft to turbojets. Hydrogen in transport, electricity generation and battery replacement technology can be expected to show some trends of the transitions reflected in the cases mentioned above.

In the following, the reader will be given insight into many of the above mentioned questions by showing how a number of successful research and demonstration projects performed around the world have tackled the many socioeconomic aspects and questions. Safety aspects will be examined, then codes and standards, and lastly some insight will be given into ongoing or newly completed projects around the world that have in common to address the question of introducing hydrogen into society.

Infrastructure

Hydrogen infrastructure refers to the physical links between sites where hydrogen is produced and where it is consumed or utilised. In this sense, hydrogen infrastructure includes pipelines over short and long distances, road transport by trucks or by rail or waterways as well as large hydrogen storage facilities and filling stations.

Over most of the past century a complicated hydrocarbon infrastructure was built all around the world. We only need to visit our local refuelling station to get in touch with such a structure. When changes are foreseen in the major energy carriers or sources, inevitably the infrastructure has to be developed or changed. This difficult challenge has been met by a variety of research work. Recalling the first chapters of this book, one can see that there is a fabric of possibilities to create infrastructure. Hydrogen has to be produced from primary energy sources. These primary sources can involve carbon capture and/or sequestration. The produced hydrogen has to be supplied to the end user. Although a gradual penetration of hydrogen into the existing infrastruc-

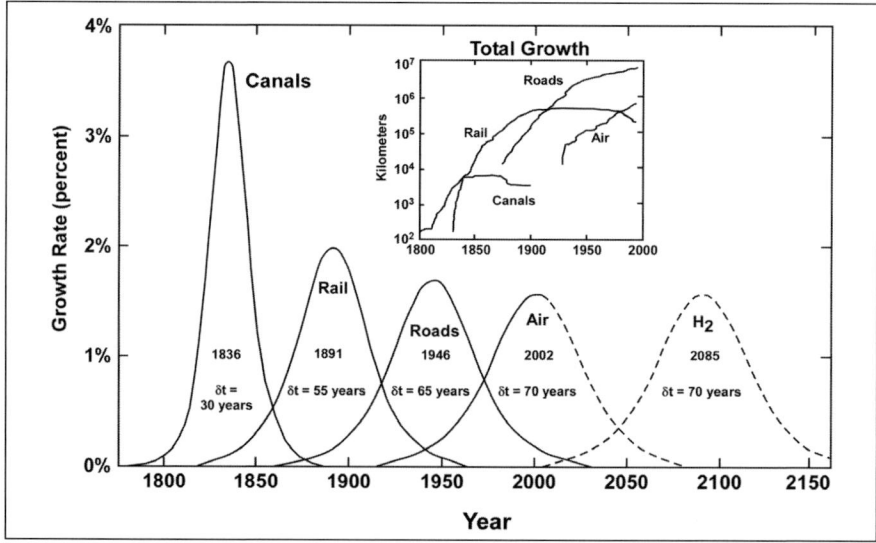

Fig. 50. Penetration of major transportation infrastructures with growth rates. An estimated scenario showing smoothed historic rates of growth (solid lines) of the major components of the US transport infrastructure and conjectures (dashed lines) based on constant dynamics. The inset shows the actual growth, which eventually became negative for canals and rail as routes were closed. Delta t is the time for the system to grow from 10 per cent to 90 per cent of its extent. We present this conceptual figure by adding hydrogen energy infrastructure into the original work of Ausubel, Marchetti and Meyer 1998.

ture is manageable, the competition from the existing pathways, availability and prices of hydrocarbons is seen by many as an insurmountable hurdle.

The production options will depend to a large extent upon local or regional conditions as we discussed in the third section. During the initial steps the new hydrogen infrastructure may depend upon the further development of the already existing electricity or natural gas network available.

In countries with ample electricity it may become favourable to electrolyse water at the fuelling station sites. In these cases the individual fuelling stations would be equipped with electrolysis systems, compression facilities and storage facilities. Dispensers have to be available and there are visions of small scale dispensers available in residential quarters or even in private residences, ready to deliver hydrogen from an electrolysis pathway.

Where natural gas is accessible, plants or units for reforming could be made available and the filling stations could be reformer plants. As in all infrastructure plans, economies of scale may require larger reformer plants and a distribution system for the produced hydrogen. Again here, the use of piped hydrogen or truck transport of liquefied hydrogen are among the options.

Among the available options, a possible pathway would be to use an existing natural gas network and distribute hydrogen mixed into the natural gas with up to 15 per cent volume proportion. This would all depend upon pipeline systems, materials, end user pressure etc. A filling station could then reform the gas or simply provide a mixture of hydrogen and natural gas as a transport fuel option. There will, however, be limitations to the possibility of transferring the existing natural gas pipeline structure for the use of hydrogen.

Firstly, when using hydrogen for combustion engines, 25 per cent by volume hydrogen content is required in order to achieve the so called optimum (Wobble index). Secondly, the density limit of 17 per cent will also affect this available option. Hydrogen being so light and diffusive will also put restraints on materials for pipelines as regards leakage, corrosion and embrittlement. When pressures go beyond some four bars, problems of the old structure could be envisaged.

Storage of hydrogen in available underground caverns is also a possibility as was discussed earlier (Tees Valley). Storage in high pressure tanks has been tested in the various demonstration projects in Europe. The pressure there is about 450 bars. The European Commission allocated ¤18.5 million to the CUTE (Clean Urban Transport for Europe) demonstration project to support nine European cities in introducing hydrogen into their public transport system : Amsterdam (Netherlands), Barcelona (Spain), Hamburg (Germany), London (United Kingdom), Luxembourg, Madrid (Spain), Porto (Portugal), Stockholm (Sweden) and Stuttgart (Germany). In Perth in Western Australia another set of three buses are being tested in a project called START. At the formal end of the CUTE project in 2006 the buses had driven over a million kilometres and some of them had been in operation over 5000 hours.

These cities want to demonstrate that hydrogen is an efficient and environmentally friendly energy carrier for the future of their cities. Twenty seven fuel-cell powered buses, running on locally produced and refilled hydrogen, should prove that zero emission public transport is possible today when ambitious political will and innovative technology are combined.

The nine European cities were convinced that the combination of a hydrogen and fuel-cell bus in a quality public transport system would lead towards the most sustainable urban transport system and address all these important problems and hurdles simultaneously.

The CUTE project was, together with the ECTOS project in Iceland, the first project world-wide which addressed at the same time the production of hydrogen, the hydrogen refilling in city centres and the operational use in commercial public transport systems. The buses have been operated on the same lines and under the same tight time schedule as commercial buses for the best comparative assessment of performance, costs and reliability.

Due to their success, the CUTE and ECTOS projects were extended for another year until end of 2006 under the name HYFLEET CUTE involving the fuel cell buses and internal combustion engine powered ICE buses. DaimlerChrysler is expecting to build their first commercial series of hydrogen buses in 2008/9 on the vast experience obtained in the European hydrogen fuel cell bus projects.

Although the technical requirements of the bus manufacturer for the refuelling station in the European bus projects were strictly the same, the resulting refuelling stations differed significantly. The same applies to the origin of the hydrogen which could be produced from electrolysis, reforming or as a chemical effluent, respectively. An EU project, HyApproval, focused on developing a handbook facilitating the approval of hydrogen refuelling stations and for assisting companies and organizations in the implementation and operations of stations.

The gas industry is actively participating in the development of hydrogen infrastructure. In Iceland dozens if not a hundred visitors from the Japanese gas industry have come to the country to watch the development of the Iceland project.

The hydrogen infrastructure demonstration activities are taking place world wide with California, Europe and Japan as main hubs.

The Spirit of Davis

Traditionally, California is one of the world cradles of the hydrogen energy society. The important movement, named California Fuel Cell Partnership, is committed to promoting fuel cell vehicle commercialisation as a means of moving towards a sustainable energy future, increasing energy efficiency and reducing or eliminating air pollution and greenhouse gas emissions.

The California Fuel Cell Partnership (CaFCP) which started in 1999 is unique and the first collaboration of auto manufacturers, energy companies, fuel cell technology companies, and government agencies. The original eight members were Ballard Power Systems, DaimlerChrysler and Ford Motor Company, BP, Shell Hydrogen, and ChevronTexaco, California Air Resources Board, and California Energy Commission.

This partnership is advancing a new vehicle technology that could move the world toward practical and affordable environmental solutions. The CaFCP has as a main goal to study infrastructure aspects and is currently consisting of 21 full members and 11 associate partners.

One of the most impressive sides of the California research on hydrogen and fuel cells is related to a very high level academic work performed at the University of California at Davis, the tranquil university town just west of the state capital Sacramento. The Institute of Transportation Studies runs an extensive so called "Hydrogen Pathways" research program studying the technical, economic, environmental and policy issues associated with using hydrogen in the transportation sector. The Davis team has contributed greatly to the knowledge-base about hydrogen infrastructure, cost and environmental characteristics of various hydrogen production pathways, hydrogen vehicle demand, and much more. Some of the interesting products that have been developed by the team include:

An impressive Compendium on Hydrogen Refuelling Equipment Costs database, edited by Jonathan Weinert, which compiles and organizes a variety of cost estimates for the major components in a hydrogen refuelling station. This includes capital costs for equipment such as compressors and storage tanks; noncapital costs for construction including design and permitting; and total costs such as cost per station and cost per kilogram of hydrogen produced. It also, to quote the Davis team "compiles actual historical cost data from existing stations and vendors. Thus, it can be used as a tool to compare existing cost estimates and to compare these estimates to real cost data".

At Davis, researchers have studied the case of coal based hydrogen production with carbon sequestration in Ohio and using fossil fuel feedstocks with the sequestration. By using low cost fuels such as coal the team argues that it can lead to low-cost hydrogen production, while carbon capture can mitigate much of the carbon dioxide emissions and lead to near zero emission and decarbonized transportation fuels. The use of coal, a carbon-intensive feedstock, will lead to the development of parallel infrastructures, one for hydrogen distribution and one for CO_2 disposal. This research will

Fig. 51 Joan Ogden visiting the Icelandic hydrogen station during a visit to Iceland.

examine possible infrastructure pathways and strategies for future hydrogen and electricity production in the United States.

One of the original approaches by the Davis team has been to examine geographic and time-dependent factors, and to understand the important underlying variables that affect infrastructure design and transitions, rather than simply carry out case studies in a few locations.

The Davis team goes even further and has modeled a hydrogen fuelling station in order to analyse the costs and environmental impacts of an "energy station" that produces both electricity and hydrogen from a single source. This work has been led by Tim Lipman. While the economics of installing a hydrogen fuelling station for just a few cars can be difficult to justify, the energy station economics are driven by its electricity generation; the hydrogen fuel is a side benefit. Lipman's model examines sites, electrical loads, number of vehicles refuelled, costs of electricity and natural gas, and a variety of international inputs.

The Davis H2 Pathways team also extends the study of infrastructure by analysing the problem by using a geographic information system. The team is using these tools to understand H2 infrastructure in a variety of innovative ways. They have developed an interactive tool for modeling hydrogen demand and methods for optimizing production, distribution, and refuelling station infrastructure based on geographic characteristics. Researchers have also begun to explore the impact of demand, centre size and number on hydrogen cost at various market penetration levels. In another study of early infrastructure in California, it was found that the number of refuelling stations needed for initial infrastructure rollout depends on the density and size of cities. Substantially less than 10 per cent of stations may be enough in very dense large cities like Los Angeles.

The research leaders, Joan Ogden, Susan Handy and Dan Sperling, point out that "the problem is the classic chicken-and-egg syndrome": the absence of hydrogen fuel stations is often cited as a major barrier to the introduction of hydrogen vehicles. Given their high construction cost, however, planners seek to build as few stations as possible while still adequately serving consumers. Recognizing that the location of stations may be more important than the number of stations, researchers developed a geographic information system model for siting generic hydrogen stations in Sacramento County, ignoring, for now, the economics of supplying those centres with hydrogen. This modeling approach provides an analytical framework for siting early hydrogen fuel stations. Initial results suggest a few strategically sited stations could be sufficient to satisfy a large number of prospective consumers.

The depth of the Davis approach is clearly visible in the development of Transitional Hydrogen Economy Replacement Model (THERM) and the Steady State City Hydrogen System Model (SSCHSM) where the researchers look at scenarios for the transition from distributed to centralized hydrogen production. Natural gas through distributed and centralised pathways is investigated and it is argued that it provides a useful point of comparison and appears to be among the most likely near-term hydrogen production options. This analysis is formulated such that the distributed on-site natural gas steam methane reformers (SMR) are eventually replaced by centralised natural gas SMR with pipeline distribution. Models are used to estimate the infrastructure transition costs as a function of relatively small number of parameters for various demand scenarios. The analysis investigates how costs and timing depend on factors such as the size and geographic density of demand, and the market penetration rate. In the SSCHSM – Pathways, researchers developed an Excel-based model that estimates the steady state cost and low-cost pathway for 73 of the largest cities in the United States. The main focus of the modeling effort is to provide the program structure within Excel to compare costs between alternative pathways, costs and cities. It provides an estimate of the infrastructure requirements and comparative costs based upon differences in city size and population, regional feedstock prices and the current electric grid characteristics.

As we have seen throughout this book, hydrogen can be made from a large variety of feedstock. The Davis team has analyzed strategies for co-production of hydrogen and electricity as well as opportunities and challenges associated with a convergence in the transportation fuels and electricity sectors. The work of the team provides an invaluable contribution for future transitions.

Joan Ogden, Obadiah Bartholomy, Andrew Burke and their colleagues together with the Sacramento Municipal Utility District are also studying hydrogen produced from renewable electricity sources a frequently quoted long-term goal for the hydrogen economy. This study examines the technical and economic realities of using wind power to produce hydrogen on a large scale in the state of California. A techno-economic model is applied to major wind resources and electricity utility demand profiles to investigate the technical, economic and environmental impacts of wind usage with respect to system design and control strategies.

Hydrogen Pathways researchers, Nathan Parker and others, have also developed engineering-economic models to estimate the delivered cost of hydrogen produced from agricultural residues using real-world data on the location and size of the residue resource and potential hydrogen demand centres. Hydrogen from this renewable resource was found to cost USD 3/kg at the refuelling station.

Antoine Simonnet, an expert at the TOTAL energy company of France, has been examining with the Davis team the technical options for distributed hydrogen fuelling stations in a market driven situation. A Davis report on the subject studies real demand profiles at TOTAL Corporation refuelling stations, as well as real electricity loads. The demand profiles are subsequently transferred to hydrogen stations and a study performed how this might affect the technology choices. The reformer technology, since it has a limit in capacity (in terms of kg H_2 produced/hour), is less adapted to unsteady demand than the liquid hydrogen station. The study quantifies these tradeoffs. This project has a French - U.S. dimension as the study also assesses the energy station in France, where a fuel cell is used to energize the station.

The Hydrogen Pathways program also considers the policy and business strategy issues associated with a potential transition to hydrogen. One such task is a long-term study of the policy process. General theories of the policy process are used to guide this study. Data has been collected from all the stakeholder sectors and qualitative and quantitative analyses performed. Initial results show that all sectors perceive oil dependency as an important problem, followed by climate change. Hydrogen is seen as playing an important role in addressing these problems. Stakeholders' policy beliefs and preferences were also obtained. For example, first priority is given to policies that support R&D on hydrogen technologies.

The program includes the participation of 15 graduate students and more than 15 faculty and research staff. The co-directors are Joan Ogden and Dan Sperling and the associate director is Anthony Eggert.

I have myself visited the Davis facility and Joan Ogden has been instrumental in helping us in Iceland by sharing with us the enormous knowledge base developed in California as well as her previous work in Princeton. The public private cooperation in California is very remarkable. The Davis projects as well as the Fuel Cell Partnership enjoy very good relations with Governor Schwarzenegger and key environmental agencies in California – a relationship dating back to an important initial support by Gray Davis. It is of importance for me when writing about the "taming of the proton" to draw the world´s attention to the impressive work performed by the dedicated team in California.

The HyWays of the European roadmap

The most advanced hydrogen roadmapping activity that has been performed in Europe is called HyWays. Three dozens of companies, institutions and universities have joined forces in this remarkable work which has been initiated and coordinated by Ludwig-Bölkow Systemtechnik of Ottobrunn in Germany.

What has been particularly impressive about HyWays is that the project has attempted to reflect real life conditions by taking into account both technological and also country-specific institutional, geographic and socioeconomic barriers and opportunities. In the project, both mobile and stationary applications are being studied, including possible synergies and spill-over effects of the two.

HyWays systematically describes the future steps to be taken for large scale introduction of hydrogen as an energy carrier in the transport and power market and notably as a storage medium for renewable energy. The Roadmap thus produced includes an action plan for the support of the introduction of hydrogen technologies.

In the first phase of HyWays, the introduction of hydrogen in France, Germany, Greece, Italy, the Netherlands and Norway has been assessed. This work included national stakeholder´s workshops and addressed energy chains and preferable pathways of hydrogen in Europe.

One of the main assumptions of the HyWays work is based on an accepted "European Energy and Transport Trends for 2030" publication already available. The team then made assumptions for the demand for passenger car transport, as well as growth of stationary fuel cells. For the analysis it proved necessary to introduce four hydrogen scenarios ranging from "business as usual" to much more hydrogen oriented development.

In the mobile scenario, demonstration projects are expected to characterise the development until 2010, from which year limited series produced vehicles are expected on the market in a first generation. From 2020, a second generation of hydrogen vehicles is expected with on board hydrogen storage and low-cost elevated temperature fuel cells. In the decade from 2030 to 2040 fuel cell powered automobiles are expected to dominate the market. This analysis leads to an estimated penetration of hydrogen fuelled vehicles in a range from 22 to some 54 per cent of the total vehicle fleet in 2040 depending on penetration of hydrogen in society.

In the stationary sector, fuel cells are expected to have a penetration which is treated differently for residential versus commercial use for combined heat and power. In the high penetration 2040 scenario, some 2.7 per cent to about eight per cent of total share of fuel cells in commercial sector and residential sector are expected respectively. It should be noted here that the HyWays assumption about origin of hydrogen produced, focuses on hydrogen from fossil fuels until 2030 with renewable hydrogen being introduced in increasing rates thereafter. The analysis has concentrated on this period and with larger uncertainties until 2050, the latter period needed to understand potential gradients for growth in the year 2030.

Of particular interest in the HyWays study is the possible future for competitiveness of hydrogen technology under assumption of various economic as well as employment profiles for Europe. This ranges from an optimistic scenario of a European lead market with export of the technology; to a more pessimistic scenario of lead markets outside the EU and the consequent imports of hydrogen technologies into Europe. The optimistic scenario is seen to increase net employment in Europe by up to 0.6 per cent by 2030. The pessimistic scenario leads to a negative, -0.4 per cent reduction of employment in the same time frame. The difference can sum up for the six countries investigated to about $^3/_4$ million new employments. GDP development as well as emission impacts are studied. National developments are studied for six countries involved and, as a whole, the HyWays report provides a deep and clear insight into the various pathways the hydrogen economy can take in the different countries.

In its second phase, which ended by June 2007, HyWays has added four further European member states to its member state portfolio, now also including Eastern Eu-

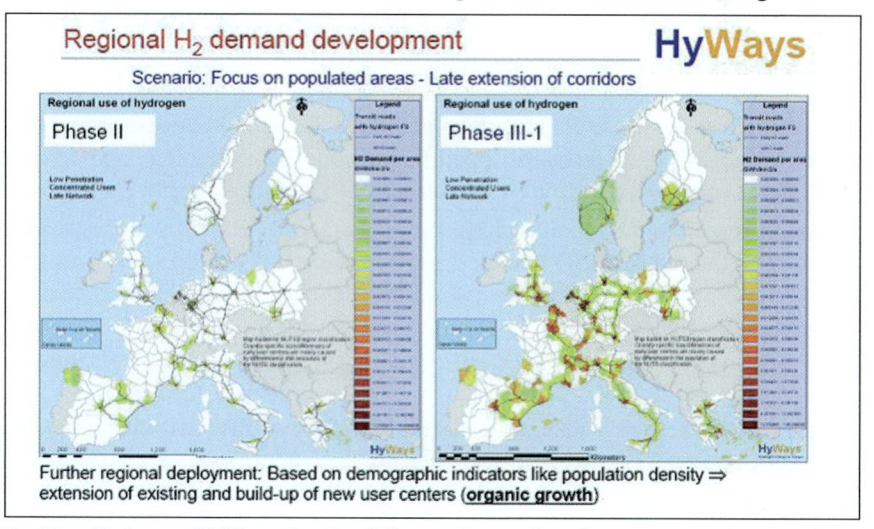

Fig.52a. Hydrogen HyWays plan for different phases of implementation in Europe. Source: LbST.

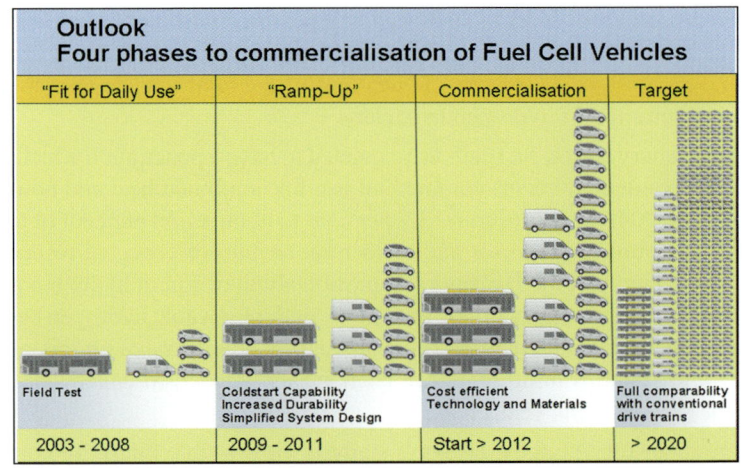

Fig 52.b The development of hydrogen fleets as seen by Daimler. Source W. Dönitz.

rope (Poland) next to Finland, Spain and the UK. Together they cover 85 per cent of the European population and 72 per cent of its land area. A number of issues will be raised for all 10 member or associate states which could not be assessed in Phase I. These include the specific prospectives of renewable energy, the expected potential of carbon capture and storage (CCS) options to provide hydrogen, as well as more stringent energy market limits (rapidly increasing fossil energy prices and CO_2 emission constraints).

Finally, the individual member/associate state results will be forged into a joint European Hydrogen Roadmap with the Action Plan as the final derivative.

The rising sun of Tokyo Bay

Japan has great ambitions towards the introduction of hydrogen and fuel cells into society. The timeline has assumed that the period up to 2005 was defined as a phase for preparing the infrastructure with demonstration programmes. The period up to 2010 has been defined as the phase for introduction of the new technology and the time up to 2020 correspondingly defined as the time for spreading the technology involving fuel supply systems and cost reduction as well as outreach among the general public.

As an example of the ambitious Japanese aims, some 50,000 vehicles were expected to be on the roads in 2010 and the number was expected to rise to 5 million in 2020 and 15 million by 2030 with the appropriate development in the number of fuelling stations.

The Japan Hydrogen and Fuel Cell Project JHFC is led by Japan's Ministry of Economy Trade and Industry and the demonstration study has been directed by the Japan Automobile Research Institute, whereas the study of the performances of fuelling stations has been led by The Engineering Advancement Association of Japan.

In the project, over 20 automobile manufacturers and energy companies are partnering to provide hydrogen produced from a variety of resources for fuel cell vehicles operating throughout Japan. The project involves operation, data analysis and a

Fig 53 Japanese Hydrogen fueling station.

few main working groups on hydrogen station, liquid hydrogen technology, public relations, safety promotion as well as total efficiency studies. Toyota, Honda, Hino, Mitsubishi and Nissan are actively participating in the project, in cooperation with DaimlerChrysler and General Motors. Seven oil and gas companies participate. Ten stations have been built in the Tokyo Bay area (see Figure 79)..

The second important demonstration experiment is the Partnership between the Fuel Cell Commercialisation Conference of Japan (FCCJ), a voluntary industry group, and the Policy Study Group for Fuel Cell Commercialisation, a government R&D group formed to meet the (We-NET) objectives. The work aims at accelerating the commercialisation of fuel cell vehicles.

The JHFC project has involved a park of the same name where visitors have been able to come to seminars and events. During the summer and winter holidays parents and students have been able to enjoy special seminars where hydrogen technology has been introduced. Nine schools in the metropolitan area have been arranging special "hydrogen for kids" programmes. The approach of the Japanese to educate children in the area of hydrogen is very deep reaching and shows their seriousness about preparing for the future.

At the beginning of 2006 the total mileage of the project stood at around half a million kilometres. The Japanese management has issued some interesting data about efficiencies of the hydrogen production. In this way the efficiencies from feedstock to fuel tank on the vehicle has been found to range from 58.7 per cent (using the Lower Heating Value definition LHV discussed earlier) in Yokohama and Senju where the feedstock for reformation has been petrol and liquid propane gas – to as high as 65 per cent (LHV) in the Kawasaki station where methanol has been the feedstock.

High stakes in a HySociety

A consortium of 20 organisations from 14 European countries undertook a project under the energy subprogramme of the the Fifth Framework programme of the EU. The programme, named HySociety, ran for two years and ended in 2005. A close cooperation was made with the ten candidate countries scheduled to join the EU in 2004 as well as Iceland and Norway.

This courageous group embedded in their vision all the "vertical issues" such as hydrogen production, storage, distribution and application, seem to stream down into society as end-user, with a number of horizontal concerns such as safety, codes and standards, and the environmental impact. In between the vertical and horizontal streams the team identified a turbulent flow with ill-defined barriers. It was exactly these barriers they wanted to tackle. One of the tasks of the HySociety group was then to take the actions present in the various world wide demonstration projects to fruition by assessing the non-technical barriers that could confront the introduction of hydrogen as an energy carrier in European society. Questions like the impact of hydrogen on society, the codes, standards and the cost of the new infrastructure, public safety concerns as well as public perception of hydrogen are being examined. The initial focus was to be able to propose a strategy to accelerate hydrogen implementation and widespread utilization.

Among the hurdles identified was a number at administration level. In some countries many agencies would be involved in the permitting procedure for a hydrogen facility; in other countries the process is simple but yet strict as the term hazardous has been given to the substance.

Through the work of Icelandic New Energy and Maria Maack, I was able to follow the development of the HySociety group through the end of the work in 2005. In their report to the Commission the group presented a so called balanced deployment strategy where technology development pace went hand in hand with hydrogen demand. The group´s careful analysis did not result in one optimal hydrogen supply chain and called for the urgent need for carbon capture to obtain lower emissions accompanying the most likely fossil based routes for hydrogen production. The group concluded that a highly carbonaceous production path offered lower emissions with the use of carbon capture and sequestration. At the same time the group pointed out that natural gas already offered a substantial mitigating potential by itself.

The report suggests an action plan for European hydrogen infrastructure in three sectors. The transport sector is urged to introduce the more efficient and less polluting fuel cell vehicles, and in fact initially, also ICE hydrogen vehicles, despite their known NOx emissions.

The Electric Power application sector is expected to introduce hydrogen for energy services that require energy storage, with special emphasis on island societies, and for some special applications such as auxiliary power units or uninterruptible power units (APU or UPS etc.).

As regards the power sector, large central power plants are expected to capture and store CO_2 independently. Here, coproduction of hydrogen and electricity would become a part of the infrastructure.

The HySociety group presents two approaches to the hydrogen infrastructure future. One for 2010 where, as regards transport, the group visualises two major regions developing infrastructure for transport with up to 250 vehicles, both fuel cell and internal combustion engine driven cars, and buses with up to ten fuelling stations.

In the years from 2010 to 2030 the group anticipates progress in carbon capture and sequestration most likely to start in off-shore CO_2 storage sites and more progress in hydrogen storage systems.

Following these developments the group sees a 2030 scenario where 85 million vehicles powered by hydrogen fuel cells, the equivalent of 20 per cent coverage and over 3,800 fuel cell based power plants. Fuel distribution is seen performed by trucks, requiring over six thousand of them, a pipeline system over half a million kilometres and some 54,000 refuelling stations. To make this hydrogen, totalling about 90 million tons annually (and about doubling the present production capacity of 50m), the HySociety group foresees 420 million tons of CO_2 captured annually with nearly 4,000 biomass gasifiers and electrolysers, over 1,000 central plants and the same number of carbon sequestration plants.

The HySociety group does not leave the subject without carefully spelling out the anticipated barriers for infrastructure and transport in the short term. They point out the fact that regulations regarding space and free space around filling stations are not yet established and that site permits will be delayed by local authorities, partly caused by some lack of public acceptance as has been seen in a number of areas. The group expresses worries about the shortage of prenormative research into CO_2 storage in geological structures and not the least the slow progress of the start up of serial production of hydrogen vehicles.

A more detailed, long term barrier view, expects problems in truck delivery of hydrogen and continued scientific scepticism on the viability of underground or off-shore storage of CO_2. This also addresses conflict of interest between national states when production and transport of hydrogen increases. The group calls for more interest among the European energy industry for hydrogen which is expected to arise from the above mentioned uncertainties.

The analysis of barriers for infrastructure in the power sector leads to similar conclusions.

Hydrogen with Foresight

The Nordic countries have a long tradition for cooperation in various fields including energy. In a Nordic Hydrogen Energy Foresight project performed 2003-2005 a Nordic group of 16 Nordic organisations, including R&D institutes, energy companies, industry and public associations tried to look ahead to examine what was needed to build the Nordic Research and Innovation area in hydrogen. The objective of the group was to develop socio-technical vision for a future hydrogen economy and explore the pathways open to commercialisation of hydrogen energy in all walks of life. In addition, the group wanted to contribute decision support for companies, research institutes and public authorities for policy development. Finally the group aimed at developing and strengthening scientific and industrial networks. The foresight process was managed and facilitated by a team of specialists in energy systems and foresight from Denmark (Risoe National Laboratory), Finland (the VTT institute) and Sweden (FOI Swedish Defence Research Agency). Interaction between research, industry and government, and combination of judgemental and formal procedures were essential features of the foresight project. The process included a series of pre-structured interactive workshops supported by systems analysis and assessment of technical developments.

I had the pleasure and privilege of working with this group as a part of Iceland's participation within the framework of the Nordic Energy Research which is a branch of the Nordic Council of Ministers cooperation portfolios.

The Nordic Foresight group defined among other things external scenarios for Nordic hydrogen energy introduction. These scenarios are quite illuminating and in many ways independent of the Nordic point of view: In a two period approach the Nordic group saw a first period setting a socioeconomic stage for a second period scenario focusing on major energy-related challenges.

In the first period of external scenarios the first one is called B-**Big business is Back**. In this scenario the Nordic group sees a globalised economy dominated by multinationals and big business motivated policy approaches with very limited interest for global environmental issues. In this bleak scenario it would, however, be expected that hydrogen energy had a far future role in energy systems.

In a somewhat more optimistic second scenario, E-**Energy Entrepreneurs and Smart Policies** were dominating in a globalised economy controlled by entrepreneurs and venture capitalists. In this scenario the energy sector is seen characterised by a tendency towards decentralisation. The interest in global environmental issues is moderate which also applies to oil prices.

The final scenario presented by the Nordic group is called P-**Primacy of Policies**. This scenario foresees the Europe-centric economy characterised by intergovernmental cooperation and a close cooperation with big business resulting in large demonstration projects in the transport and energy sector. Global environmental issues enjoy some interest under the shadow of a high oil price level which is kindled by security-of-supply problems which in turn result from high oil prices driving the urge for an energy sector change.

When the group started to rate the various pathways it became clear that the share of hydrogen in the Nordic scenario of 2030 could be very different. The lowest share of hydrogen was found in a B-Big business scenario with only six per cent while a much higher proportion would be found in the E-Energy Entrepreneurs and Smart Policies scenario dominated by hydrocarbon insecurity of supply – resulting in 15 per cent share in 2030. The most impressive scenario for hydrogen introduction, the P-Primary of Policies scenario, was a natural winner with 18 per cent share in this foreseen future world suffering from undisputable CO_2 problems. The percentage in the scenarios excluded the energy intensive Nordic industry sector.

The Nordic work is of great interest and can provide knowledge to anyone interested in creating future scenarios and strengthening the knowledge creation and strategic intelligence – resulting in best business options and pathways for a hydrogen economy.

The concluding recommendation from the Foresight work is to call for information and awareness campaigns on the hydrogen economy and innovation in a closer cooperation among the countries involved. The need for continued demonstration projects is seen as vital as is the stimulation of niche markets. International cooperation is seen as a must.

Hydrogen Education

As the hydrogen economy emerges on the horizon, education becomes crucial to the global transition, providing political leaders, technical specialists, and laypeople with the knowledge necessary to play their own, appropriate roles in the transition to a hydrogen-based energy infrastructure.

Education is a crosscutting issue and touches upon almost all levels and individuals in society. There is for example the need to inform and educate all government officials ranging from national to local; the safety and code officials; the university and college students as well as primary and secondary teachers and students in actions reaching out to the general public.

The identified audience groups each have distinct hydrogen educational needs ranging from general to technical, broad-based to narrowly-targeted. Certain audiences require special attention in the near-term to facilitate research, development and demonstration efforts, while the needs of others will grow over time and as the transition to a hydrogen economy progresses. In some of the existing demonstration projects in Europe, North America and Japan, the opportunity to educate and train a variety of human capital in the field of hydrogen has been seized with clear successes in building a knowledge base. In some countries and territories the effect of these demonstration projects will result in a public outreach and education of important value.

In some universities, strong emphasis on hydrogen has been manifested with a resulting "centre of excellence" created. Many of today´s corporations possess enormous skills in hydrogen and enjoy workforces of hundreds of knowledgeable hydrogen manpower.

In some cases, regional authorities have joined forces to make an impact in hydro-

Fig. 54 WBU, the headquarters of Hydrogen education in Ulm, Germany.

gen education. An outstanding example is in Ulm, the birthplace of Albert Einstein in Germany, where a fuel cell education and training centre (Weiterbildungszentrum Brennstoffzelle Ulm) was inaugurated in 2005 (see Figure 54)..

The aim of the centre is to provide high level education and training in the field of fuel cell and hydrogen technology addressing the needs all the different target groups from trade, industry, academia, and schools. The centre enjoys a five year support from the state of Baden-Württemberg and the Federal Ministry of Economy and Labour. This initial support has made possible the construction of a building devoted to education and training and the initial provision of fuel cell hardware. Experimental systems were provided by the German Aerospace Centre (DLR), the Fraunhofer institute for solar energy systems (FhG-ISE) technologies and the solar and hydrogen energy research centre (ZSW). The technical content of the education and training programs will be jointly developed with Forschungszentrum Juelich (FZJ) and representatives of trade and industry. This is a truly remarkable public-private achievement taking place in Germany. The Hydrogen and Fuel Cell Network in North Rhine – Westphalia is the most powerful of all the German networks.

Also within the realms of the United Nations, hydrogen has been gaining attention, a trend reflected in the establishment of The International Centre for Hydrogen Energy Technologies (UNIDO-ICHET) in October 2003. The centre is the result of an agreement between UNIDO and the Turkish Ministry of Energy and Natural Resources. Under the terms of this agreement, the centre will act as a conduit for knowledge and technology flow between the developed and developing nations through the provision of support, facilities and expertise concerning all aspects of energy conversion technologies involving hydrogen.

The Centre, initiated by Professor Nejat Veziroglu, will be actively involved in communicating the latest technologies to those groups and organizations, particularly those in developing countries, that can benefit most from the tangible advantages the use of clean and renewable energy sources can bring. From its base in Istanbul, UNIDO-ICHET will provide educational, training and applied research programs to visiting scientists and engineers, and expertise and support for the establishment of industrial scale pilot studies at selected locations across the world.

Sophisticated Economics of Hydrogen Energy

The economics of hydrogen energy is perhaps the most difficult research and development aspects we are faced with. On one hand, economic analysis has to look at the anticipated growth in old and new markets for hydrogen. On the other hand, they have to involve assessment of all sorts of externalities and other market distorting influences of hydrogen based technologies. Externality costs can have many causes. As Anthony Owen, President for International Association for Energy Economists, has argued that if sustainable development and energy security can be regarded as public goods, then their level of provision through competitive market forces would be sub-optimal and this would justify market intervention designed to raise their supply to a level that would be optimal to society. Then, as he sees it, the hydrogen energy economy would be one option for addressing this situation.

Hydrogen energy economy involves many new technological and ecological concepts. Platinum and other catalysts are expensive and scarce. Using natural gas for the production of hydrogen would again raise the demand for the gas which would increase both the price of natural gas as well as the hydrogen. This would put pressure on more drilling for gas and better utilisation. We have seen how natural gas is replacing coal for electricity production, so a competition base with hydrogen production would be created.

A transition should not be expected to happen in a short period. History shows that transitional periods are characterised by long times of coexistence of the old and the new technology. Coal replaced wood; oil replaced coal and the steam engine substituted waterwheels over a relative long period of transitional time. Better understanding of fluid dynamics would improve waterwheels and that would take place almost simultaneously with the use of the steam engine. This was perhaps best seen in the invention of Victor Poncelet of new curved water blades which improved the efficiency of waterwheels. And we could name a variety of examples showing how history happens in cycles.

A similar effect is seen as hybrids replace conventional petrol cars on the roads. Efficiency is improved, even when using the old fossil fuel. Hybrid development is a prerequisite for better utilisation of hydrogen and fuel cells in transport and will undoubtedly pave the way for hydrogen. We would generally expect simultaneous development of new and mature technologies and therefore gradual transition. Coexistence of petrol/diesel fuelled cars and hydrogen cars would be expected long into the future in some places.

The development of infrastructure provides another challenge for the hydrogen economy. The chicken-and-egg problem of fuelling station versus hydrogen cars is a well known dilemma in the development of the hydrogen society. When the history of the petrol powered automobile in the United States is examined, it becomes clear how important it was to develop the roadside petrol stations connecting the population density points and spreading the automobile to finally "inhabit" the whole of the U.S. To make this possible, enormous public financing of infrastructure was indeed necessary.

The various demonstration projects available for assessment of competing fuels in the context of transportation and stationary power generation sectors, can give some basis for evaluating the competitiveness of hydrogen. Most demonstration projects are done with prototypes or series of prototypes and are only capable of reflecting roughly the anticipated economical factors.

For such assumptions, mid-term to long-term estimates of prices and availability of fossil fuels, their availability etc. have to be made, as well as cost of water and various technologies. Many aspects have to be borne in mind: investment expenses and cost of the R&D phase, technological breakthroughs, scaling effects and success with carbon sequestration technology – as well as the decisive status of the global economy two or more decades into the future.

Projected hydrogen demand varies significantly in the studies of different countries. In an International Energy Agency (IEA) study, Canada expected an annual 2.7 per cent growth rate in hydrogen demand; whereas the Swiss expected up to 7.7 per cent increase.

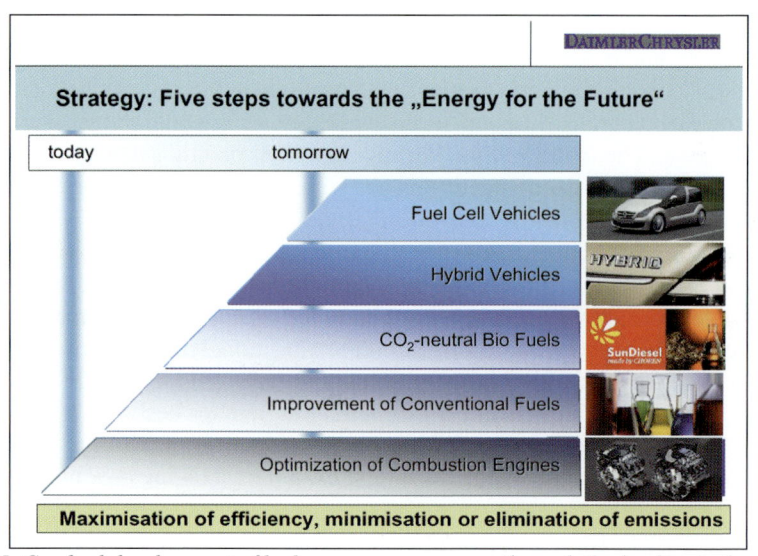

Fig.55 Gradual development of hydrogen economy seen through the leading technology. Source Daimler, W. Dönitz.

If we compare the studies and scenarios from the various studies made for 2025, we note that the projected average production increase amounts to up to 20-25 per cent in for example transport. The results from most other studies are not very far from this result, but suffer from a lot of uncertainties as we have described in the sections above.

Finally it can not be overemphasised that building infrastructure requires long time. Studies of infrastructure formation shown in figure 50 indicate how long it took on average to build and perfect different transport infrastructure backbones: canals, rails, roads and air traffic. The average time for each technology spans from one to nearly two centuries. Expecting a mature infrastructure for hydrogen to be built this present century seems quite natural. This may, however, require a political will as was reflected in President Eisenhower's urge to build the automobile infrastructure of the United States in the post World War II period.

Rules of the game: Hydrogen safety, codes and standards

Tampering with the elements water and fire

First a little digression. In 1973 when I was 19 years old, my hometown of Vestmannaeyjar experienced a volcanic eruption when a fissure on the main island of Heimaey – one of a group of volcanic islands south of Iceland - started to erupt. The whole event did not result in any natural casualties, but for a time the lava streaming from the new volcano threatened the excellent natural harbour on Heimaey island. This harbour was the lifeline of a thriving fisheries economy in Vestmannaeyjar. In its heydays in the 1960s it had been responsible for up to 12 per cent of the export income for Iceland.

One of my mentors, Professor Thorbjorn Sigurgeirsson of the University of Iceland had tested ocean water cooling of lava flows in the newly born island of Surtsey a

Fig. 56 Seawater sprayed on a volcanic lava front in Vestmannaeyjar town. Source: Sigurgeir Jonasson.

decade earlier, and was the initiator of an extensive cooling of the lava running out of the new crater and threatening the harbour on the island. The competition between Nature and the engineering team seemed for a time to be doomed to be lost and the lava moved closer day by day, swallowing and digesting all infrastructure it hit on its way.

I remember that Thorbjorn told me about a brilliant American scientist who suggested to him to ask the NATO defence airforce from Keflavik Iceland to administer an explosive load to the crater side facing away from the harbour in order to divert the lava flow in a direction opposite the harbour and thus save Vestmannaeyjar harbour.

Thorbjorn told me that after a careful analysis they decided not to perform the operation. The danger of ocean water rushing into the crater following such an explosion could have meant conditions where water and hot molten lava had mixed in a deadly mixture. The steam produced from the interaction of hot lava and seawater contains an abundance of toxic hydrochloric, HCl, gas. This acidic mixture of water vapor and HCl is known as *laze* and is well known in Hawaii.

What is even more important, hot lava and water, in the extreme case of a collapsing seaside crater, could split a part of the water and create hydrogen and oxygen like in the well known steam-iron process. This hydrogen, in turn, could in principle fuel an explosion that would have become comparable to the infamous Krakatoa explosion in August 1883. Fortunately, by spraying seawater on the lava in Heimaey, the team succeeded in stopping the lava flow down to the harbour side and save the harbour.

The unexpected steam-iron process

About two decades later I was reminded about these enormous powers of Nature when working with the task of renewing the cooling system of the Icelandic Alloys Ferrosilicon plant in Hvalfjördur, Iceland. I had devised a spray water cooling in order to cool the 1600°C glowing hot ferrosilicon as fast as possible to create a stronger finished product which they called Strongsil. Remembering the Heimaey story, I had told the team at the Icelandic Alloys plant never to mix the hot alloy and water more than was necessary with the spray cooling.

But one day it happened: A former sailor, who loved the sea, was working on a night shift, transferring extremely hot wagonloads of ferrosilicon to a storage site for further cooling. Without thinking about the consequences, or just because as a sailor he loved water, he had picked a cooling box from the storage site outside the plant house and brought it inside. The box contained about a foot of fresh snow from a recent snowfall. The "sailor" prepared the box and then tipped a 1200°C one ton loaf of metal into the box: Bang!

Fortunately, no one was hurt. I was called the morning after and the Icelandic Alloys team described a very powerful explosion. This came as no surprise. However, what surprised me was the description of what seemed a secondary explosion up near the rooftop of the plant. This blast apparently had thrown some square metres of corrugated iron roof outside and left a large hole in the roof.

A careful analysis showed that the good old sailor had exactly done what was not desired: he had mixed water and molten alloy. But the suspicious secondary explosion was initially a mystery. Further analysis would solve all questions. What probably happened was that some free hydrogen, probably formed in a so called steam-iron process when the water spray cooling droplets hit the surface of the metal, had been forming a "cloud" near the inside of the rooftop, had exploded when glowing alloy bullets hit it a second earlier. The result was a powerful hydrogen related explosion. Hydrogen had reacted with atmospheric oxygen and exploded with the consequences previously de-

Fig 57. Ferrosilicon spraycooled in the Grundartangi plant.

scribed. Hydrogen is highly flammable. Let us look deeper into some of the consequences of this flammability and related aspects.

The annual world production of hydrogen is currently exceeding fifty million metric tons with an energy content close to 5,700 PJoule. Traditionally, the hydrogen industry is has been run by very experienced engineers. This means that the industry has from the very beginning developed strict procedures when using hydrogen. This dates back from the earliest coal based town gas system in Paris in 1837. From the experience of the professional handling of hydrogen through the years, and by comparing it with other fuels, it could be stated that the safety record is surprisingly good. From our own experience in Iceland, we had since after World War II a Fertilizer plant producing fertilizers for the volcanic soils of the sub arctic island. The production involved about two thousand tons of electrolytic hydrogen annually for about 60 years. From my discussion with the leading engineer of that time, Mr. Runolfur Thordarson, the operation was close to trouble free for about half a century, with the exception of problems with ammonia, which is made from the hydrogen. In the end the Fertilizer plant had to give up because of competition from imported, cheaper fossil energy based fertilizers. But Thordarson adds that "hydrogen's power has to be respected". I have given these two extreme examples from Iceland to show how important knowledge of hydrogen is and that it is a determining factor for safe conduct.

Hydrogen safety

Now, as the world turns much closer attention to commercialised hydrogen and as more inexperienced people will handle it in the form of commodity, it becomes ever more important to exercise caution when dealing with the enormous chemical energy associated with this element.

Let us for a moment take a look at **table IV 1** which shows the main safety-related properties of hydrogen compared with methane (natural gas) and petrol. First let us concentrate on the diffusion coefficient in air, the measure which determines for example how quickly a leaking pipe will disperse hydrogen to its surroundings. The diffusion coefficient for hydrogen is about four times larger than for methane and twelve times larger than that of petrol. In the case of a leak close to a fire or a hot point source, hydrogen is likely to diffuse away whereas the heavy and dense fossil fuels could accumulate close to the source and possess more hazards if ignited. Thus, petrol, with its high density and slow dispersal, would have a tendency to congregate near the ground increasing the risk of explosion.

The diffusivity and buoyancy of hydrogen possesses problems because the element can build up in "pockets" or "clouds" in confined spaces like rooftops or the like. The buoyancy of hydrogen in air can accelerate this tendency and detection of hydrogen in eaves, or inside rooftops, is of crucial importance. In an open escape into air, hydrogen would be expected to disperse and its concentration would decrease quickly. Exactly this nature of hydrogen makes it important to administer new thinking when designing infrastructure of hydrogen and calls for new working principles.

One of the difficulties with hydrogen is associated with the invisibility of its flame in daylight. Unlike the coloured flame of fossil fuels, the flame of hydrogen can only be seen at night. On the other hand, hydrogen gives out much less heat radiation

Comparison of Hydrogen and Methane Properties with Other Common Fuels

Property	Hydrogen	Methane	Propane	Gasoline Vapour
Buoyancy (density relative to air)	0.07	0.55	1.55	3.4 - 4.0
Molecular Diffusion Coefficient (cm2/sec)	0.61	0.16	0.12	0.05
Rammabili ty range, (vol % in air) LFL – UFL	4.0 - 75	5 - 15	2.4 - 9.6	1.4 - 8
Explosive range, (vol % in air) LFL – UFL	18 - 59	5.7 - 14	2.7 - 7	1.4 - 3
Most Easily Ignitable Mixture (vol % in air)	29	9	4	2
Explosive energy (relative to H2 by vol)	1	3.5	10	22+
Adiabatic flame temperature in air (°K)	2,318	2,148 – 2,227	2,385	2,470

Source: Gary Howard, P.Eng. Stuart Energy

Table IV 1. Some properties of hydrogen.

when combusted. This is caused by the fact that the water vapour formed absorbs the heat of the burning. The overall effect is that hydrogen will cause less secondary fires due to ignition by heat radiation.

A relatively new research project HYFIRE in Europe concentrates on hydrogen combustion in the context of fire and explosion safety. The project looks at hydrogen jet flames from very high pressure release and flames impinging on surfaces. Hydrogen combustion in semi-confined and vented geometries is a part of the project as are the sophisticated "deflagration-to-detonation", DTD processes which are so important in the full understanding of hydrogen safety.

Looking back at the safety properties we notice that the autoignition temperature for hydrogen is somewhat higher than that of methane, but about twice as high as for petrol. In other words, to create a condition in which hydrogen ignites itself one needs much higher temperature than for petrol. The flammability of hydrogen as measured in air is very broad or about 4-75 per cent compared with 5.3-15 per cent for methane and 1.4- 8 for petrol. It must also be added that the lean flammability limit for hydrogen which is important for the design of hydrogen equipment, can be as low as 0.8 volume per cent.

When hydrogen is ignited in air, its speed of combustion is six times more rapid than that of methane or petrol, and its explosion speed is however in the same range as for the two fuels we are comparing with.

One more difficulty is related to the protonic nature of hydrogen This is hydrogen embrittlement of metals. When the small atom approaches a material wall, like, say inside a thin balloon, it will easily diffuse through the polymer skin and the balloon quickly drops pressure and becomes sloppy. When hydrogen is confined in a metal cylinder, the atom has the tendency to diffuse into that material. In high pressure containers or flasks, which are made of carbon containing stainless steel -with the purpose to improve tensile strength to withstand high pressures-, the hydrogen is attracted to a region close to a carbon atom in the crystal structure of the alloy. Gradually the second, third and fourth hydrogen atom accumulate around the carbon atom to form a methane molecule within the steel structure - with the inevitable consequences for the stability and strength of the steel-. Such direct physical decarbonisation of steel is a lurking problem in container design for hydrogen. The classical steels such as austenitic steel types 314 and 316 often used for various construction work have been found not to be suitable for many of the demanding tasks. Most metallic materials solidify in a polycrystalline form and the grains are held together like stones in a pavement. Hydrogen, due to its small size, diffuses in and along grain boundaries in materials. The properties of various steels for use in pipelines have been extensively studied. Designers look for high yield and tensile strength steels with toughness and weldability and in some cases the use of oxygen, carbon monoxide and sulphur dioxide has been considered to enhance the resistance of the steel to hydrogen embrittlement.

The hydrogen experts at the German Aerospace Centre (DFVLS) have also studied, in addition to various mixtures of fuel, the influence of inhibitors and geometry, detonation propagation, as well as various failure criteria for a given hydrogen container/pipe configuration. Sophisticated programmes have been developed to simulate these effects.

Generally, hydrogen has been studied very extensively with a number of container materials and its behaviour in carbon reinforced composites and austenitic steel is gradually well understood. Much knowledge is required about material behaviour before a given material can be cleared as suitable for a hydrogen container.

Hydrogen as a commercial fuel of the future calls for a whole new thinking of the safety aspects. The energy carrier will not anymore be guarded by experienced specialists in the chemical industry but will be put in the hands of ever more inexperienced people. This is a great challenge. It will require new design parameters for infrastructure, from production to transport and handling to individual hydrogen driven utilities or equipment.

Even when an infrastructure has been built around a hydrogen system, much care has to be taken with the development of static electricity in the vicinity of hydrogen. A sophisticated grounding arrangement is usually recommended for a hydrogen system to avoid static build up.

There are many misconceptions and image problems surrounding hydrogen. At a hearing within the UN in New York last year I was asked about "the combined effect of hydrogen and the risk of hydrogen bombs"! There is, of course, no linkage between the two. Hydrogen bombs are atomic bombs that use hydrogen in stead of Plutonium or some fission elements. Certainly hydrogen atomic bombs do not produce water!

We are also often reminded of the fate of the Hindenburg airship in 1937 which came down in flames after a fire and explosions. The giant airship was attempting to land in an electrical storm outside of Lakehurst in New Jersey. Witnesses reported observing a blue glow on the top of the colossus. The investigation into its causes led to a blame on the lacquer layer having burnt. The outer skin of the airship was composed of either cellulose nitrate or cellulose acetate with aluminium flakes intended to reflect sunlight and keep the skin and gasmass cool. This combination of nitrate and aluminium is commonly known today as the recipe for rocket fuel.

Whatever caused the fatal destiny of Hindenburg, I think this is not the issue: Who in his right mind would sail across the ocean on a balloon filled with explosive gas?! Hydrogen is an extremely powerful fuel and should be treated with all respect available.

Another question I often get asked relates to the fact that, unlike other gases, hydrogen warms up when left to expand. This is due to the so called Joule-Thomson isenthalpic expansion coefficient which is different for hydrogen compared to most gases which cool down upon sudden escape to lower pressures. The amount of heating when hydrogen is released from 200 bar into the ambient is of the order of 5-8 Centigrades.

Fig. 58 Hindenburg over New York early 1937. Source: Geschichte der Kölner Luftfart.

Acceptance in society through Codes and Standards

Increased safety concerns call for improved codes and standards for hydrogen. Codes and equipment standards can help to overcome industrial barriers to commercialisation and facilitate public acceptance of new hydrogen technologies. In the case of hydrogen, codes and standards are needed in a number of areas including for the vehicle and vehicle-infrastructure such as fuelling stations and pipelines. Codes and standards provide a systematic and accurate means for measuring and communicating product risk and insurability to the customer, the general public, and fire-safety certification officials.

Standards are a set of technical definitions, guidelines, and instructions for designers and manufacturers. They are are usually voluntary, but have been agreed upon to ensure consistency, compatibility, and safety. Developing a standard is a consensus process involving a number of experts in the field. Once developed, standards are usually incorporated into codes that, in turn, must be adopted by State and local jurisdictions to become legal and binding. Cyber-Appendix S describes some of the most common standards in more detail.

A task force established by the International Partnership for the Hydrogen Economy has recommended six steps to promulgate hydrogen regulations, codes and standards in the near future. The first one involves establishing a validated database of technical information on hydrogen. Second step is to develop performance-based rather than product-specific regulations, codes and standards RCS. The third to support the adoption of harmonized international regulations by for example identify existing "gaps" in the RCS work involving many international bodies. Fourthly, the IPHE recommends collection of operational data from research, development and demonstration projects and making them publicly available to a large extent. The fifth step involves education and training in codes and standards and safety for elected and appointed officials, regulators, students, users and the general public. This should be done through workshops, joint studies, summer schools etc. Finally IPHE calls for support for public and technical forums and workshops to awareness-building on hydrogen issues.

THE INTERNATIONAL HYDROGEN MOVEMENT

International Hydrogen Energy Association and the hydrogen family

The autumn of 1973 witnessed the first large oil crisis when member countries of OPEC plus Syria and Egypt announced serious restrictions on petroleum exports. A winter in turmoil was to follow. By then I was a student at the University of Copenhagen and I will never forget the sight when the Raadhuspladsen, Town Hall Square, witnessed a day without car traffic, one of the actions taken by the Danish government.

The following spring the first international conference on Hydrogen Energy was held in Miami Beach Florida,"The Hydrogen Economy Miami Energy Conference", or the THEME Conference for short, on 18–20 March 1974. The Turkish-American engineering professor, T. Nejat Veziroglu, one of the pioneers of the movement has described when, on the afternoon of the second day of the conference, a small group got together for discussing the future. This group was later named the "Hydrogen Romantics", and consisted of Cesare Marchetti, John Bockris, Tokio Ohta, Bill van Vorst, Anibal Martinez, Walter Seifritz, Hussein Abdel-Aal, Bill Escher, Bob Zweig, the late Kurt Weil, together with Veziroglu and a few other enthusiasts. There was a passionate, yet focused debate which resulted in an agreement to aim for the establishment of an international hydrogen energy association.

Figure 59. Nejat Veziroglu

In Veziroglu´s own words the discussion "turned to whether there was a need for a formal organisation. It was Anibal Martinez of Venezuela (incidentally the one who took part in setting up the petroleum cartel (OPEC), who urged the founding of a society dedicated to crusade for the establishment of what seemed to be to the gathering, and which later proved to be, the inevitable and universal energy system. It was ironic that he was proposing the establishment of an organisation, which would make OPEC obsolete. The rest is history. IAHE was established by the end of that year, and started working in earnest"

Nejat Veziroglu is quite a remarkable person. A native of Turkey, who graduated from the City and Guilds College, the Imperial College of Science and Technology, University of London, with a number of degrees in Mechanical Engineering, Advanced Studies in Engineering and Heat Transfer.

After serving in some Turkish government agencies as a Technical Consultant and Deputy Director of Steel Silos, and then heading a private company, he joined the University of Miami Engineering Faculty, and served as the Director of Graduate Studies, was Chairman of the Department of Mechanical Engineering, and the Associate Dean for Research. Until lately he was a Professor at the Mechanical Engineering Department of the University of Miami, Coral Gables, Florida, and the Director of the Clean Energy Research Institute, but has temporarily moved to Turkey where, in 2004, he established an International Centre for Hydrogen Energy Technologies under auspices of UNIDO.

One of the first activities of the International Association for Hydrogen Energy was to organize the biennial World Hydrogen Energy Conferences (WHECs) to provide a platform for the hydrogen energy community, for the scientists, energy engineers, environmentalists, decision makers, and the thinkers of the "future of humankind and the Planet Earth".

WHEC Conferences have been held in most of the major countries around the world. The first WHEC Conference was held in Miami in 1976, and the others followed in two-year intervals in Zurich, Tokyo, Pasadena, Toronto, Vienna, Moscow, Honolulu, Paris, Cocoa Beach, Stuttgart, Buenos Aires, Beijing, Montreal, Yokohama, Lyon and the next one planned in Brisbane, Australia in 2008, and then in Essen, Germany in 2010.

Another major achievement of the IAHE was the establishment of The International Journal of Hydrogen Energy (IJHE), as its official journal. In 1975, it started as a quarterly. Three years later, it became bimonthly, in 1982, was elevated to monthly, and starting in 2007 it has 18 issues per year, published by Elsevier. Over the last 30 years this journal has been the premier publication in the field of hydrogen energy.

A number of books has been published by the IAHE, and an internet site is working (www.iahe.org). The IAHE has had an enormous impact on hydrogen outreach and can be seen as one of the most important milestones in the development of a hydrogen energy economy.

The IAHE gives awards to outstanding work related to hydrogen as an energy carrier. These awards are named in honor of the following pioneers:

IAHE Jules Verne Award for Superior Service (a general area of involvement). Jules Verne, of course, predicted the hydrogen energy idea in his 1874 novel The Mysterious Island.

IAHE Rudolph A. Erren Award for Leadership in Thermochemical Area (involvement with heat engines and combustion, thermochemical production, facets of hydrogen transmission, distribution and storage, e.g., metal hydrides). Rudolph Erren was the noted dynamic German developer of hydrogen fueled motor vehicles, demonstrated in fleet service in the 1930s.

IAHE Sir William Grove Award for Leadership in Electrochemical Area (involvement with fuel cells and electrolysers, and other electro-chemical means relating to hydrogen processing). Sir William Grove was the inventor of the fuel cell in England in 1839, producing electricity as we saw in Part II.

IAHE Akira Mitsui Award for Leadership in Biological Area (relating to biological processes for producing hydrogen and synthesizing valued products utilizing hydrogen). Dr. Mitsui was a noted Japanese-American researcher in the photo-biological hydrogen production field, using special types of algae and micro-organisms to this end.

IAHE Konstantin Tsiolkovsky Award for Leadership in Aerospace Area (relating to the use of hydrogen as the leading aerospace propulsion fuel and as energy carrier in space vehicles, satellites and stations). Professor Tsiolkovsky was the Russian pioneer of Astronautics, who first proposed hydrogen-fueled rocket propulsion for spaceflight in the late 1890s.

Cyber-Appendix IAHE lists some of the main awards recipients related to IAHE

The International Energy Agency Hydrogen Implementation Agreements. The hydrogen expert assembly

The oil crisis of 1973 had many aftereffects. One of them was the creation of The International Energy Agency in 1974. The purpose of the IEA has been to bring together a broad range of experts in specific technologies to address energy-related challenges in a collaborative manner. The resulting global view and global sharing of the results are some of the main functions of the agency.

The agency is the backbone of a remarkable international movement concerning hydrogen energy. In 1977, the IEA took an important step when the Hydrogen Implementing Agreement HIA was signed over a quarter of a century ago. The HIA can take credit for the extraordinary consensus that seems to have emerged worldwide over the past quarter century. The various publications by IEA on hydrogen are accessible as published books or booklets or on the internet.

The HIA chairmen have all contributed greatly to hydrogen know-how. The first chairman of HIA was J.P. Cotzen from CEC 1977-1982 followed by J.B.Taylor of Canada, and his countryman A.K. Stuart, then W. Raldow from Sweden, G. Schrieber Switzerland, N. Rossmeissl USA, Tryggve Riis from Norway and from 2005, Nick Beck of Canada.

In the time of HIA, tasks have been chosen very carefully and approached with a high degree of professionalism. In many ways the HIA has assembled some of the best experts on Earth and resulted in increased understanding and awareness of hydrogen and its possibilities.

At the turn of the millennium about 21 of what are named "annexes" have been worked on or even completed while nine annexes have been under consideration since 2000. These are Photoelectrolytic Production, Photobiological Production, Hydrogen from Carbon-Containing Materials, Solid and Liquid State Storage, Integrated Systems Evaluation, Hydrogen Safety, Hydrogen from Waterphotolysis, Biohydrogen and lately both High and Low Temperature Production of Hydrogen.

The photoelectolytic Production of hydrogen, so called task 14, and led by Andreas Luzzi, included for example he development of the world´s first water splitting catalyst; engineering progress with WO_3 and Fe_2O_3; pioneer-manufacture of of demonstrator BEC fuel cells; and novel planar multi-junction PEC water-splitting cells (WO_3/TiO_2).

The Photobiological production of hydrogen is called taks 15. It has been completed under the firm direction of Peter Lindblad of Sweden. Of all the good things that have come out of task 15, the major breakthrough has been with the identification of accessory genes needed for assembly of the Fe hydrogenase tied to the green algae producing hydrogen from sunlight and water using photosynthesis. The other one has to do with methane and fermentation systems leading to hydrogen production.

Task 16 concerns Hydrogen from Carbon Containing Materials. This annex has led to many interesting results. A part of it has to do with large scale integrated hydrogen production and power generation, Precombustion Decarbonisation. Another part has to do with hydrogen from biomass and yet another part has to do with small-scale stationary reformers. Mrs. Elisabet Fjermestad-Hagen has led this annex.

An enormous effort has been put into task 17, Solid and Liquid State Hydrogen Storage Materials spanning about 35 projects. The leding person here has been Gary Sandrock. One of the beautiful outcomes of this annex is a database on hydride materials (www.Hydpark.ca.sandia.gov).

Integrated Systems form another annex, task 18. One part of this annex is about the development of information database and support for a Hydrogen Resources study – another one is an analysis of demonstration projects. Dr. Susan Schoenung is leading this annex.

Safety of Hydrogen forms annex 19. This involves risk management and quantitative risk analysis (QRA) and collaborative effort to evaluate the effect of equipment or system failures under real-life conditions. William Hoagland is responsible.

Finally we mention task 20, Hydrogen from Water Photolysis, a continuation of task 14 under the same leadership as the old task – the aim is still to improve the PEC materials.

The hydrogen heritage of HIA is a great contribution for the world. I have felt that a very special HIA culture has been developed over the world especially among specialists and professionals. Nevertheless, great economies like China and India, who are not members of IEA, have not been instrumental in the HIA work. This was one of the motivations for the establishment of the IPHE, the ministerial approach described in the next chapter.

The International Partnership for the Hydrogen Economy IPHE: Governments United for Hydrogen

A cool Novemberday in Washington DC, a gathering of three dozens of delegations from all parts of the world was held in the Omni Shoreham Hotel in Washington. The venue was the x hotel just west of the White House. The host was Mr. Spencer Abraham, the US Energy Secretary who greeted the guests and red a letter from President George W. Bush. The president called Fuel Cells "one of the most encouraging, innovative technologies of our era" and urged that they could be produced from resources available domestically to all the nations of the world. Bush also addressed the scientists directly, saying that by "working toghether scientists can overcome the obstacles and significantly accelerate the development of hydrogen technologies. As partners in the IPHE, we must coordinate research and development to make it possible for the first car driven by a child born today to be powered by hydrogen and pollution-free".

Over 700 delegates attended the four day meeting including 280 senior private sector representatives from hydrogen technology development, manufacturing, research, engineering and consulting firms and associations; 250 official party delegates from energy, environment and foreign ministries; 150 USG officials; 14 government officials from observer nations; and 13 credentialed media.

On November 20, Secretary of Energy Spencer Abraham led ministerial representatives from the following fifteen nations and the European Commission in the signing of the Terms of Reference (TOR), thereby creating the IPHE as an institution designed to facilitate coordinated research on emerging hydrogen technologies. As the first Secretary of IPHE, Robert Dixon later pointed out, the IPHE was a remarkable

Figure 60. Robert Dixon directing an Interagency meeting in Washington

flock of countries: the IPHE Partners' Economy span over $35 Trillion in GDP, 85 per cent of world GDP and involve nearly 3.5 billion people. Furthermore, the IPHE member countries produce over 75 per cent of electricity used worldwide but are of course responsible for 2/3 of CO_2 emissions and energy consumption of the planet.

The official purpose of IPHE was declared : "To serve as a mechanism to organize and implement effective, efficient, and focused international research, development,

Figure 61. IPHE signing ceremony in November 2003

Hydrogen Infrastructure and Society

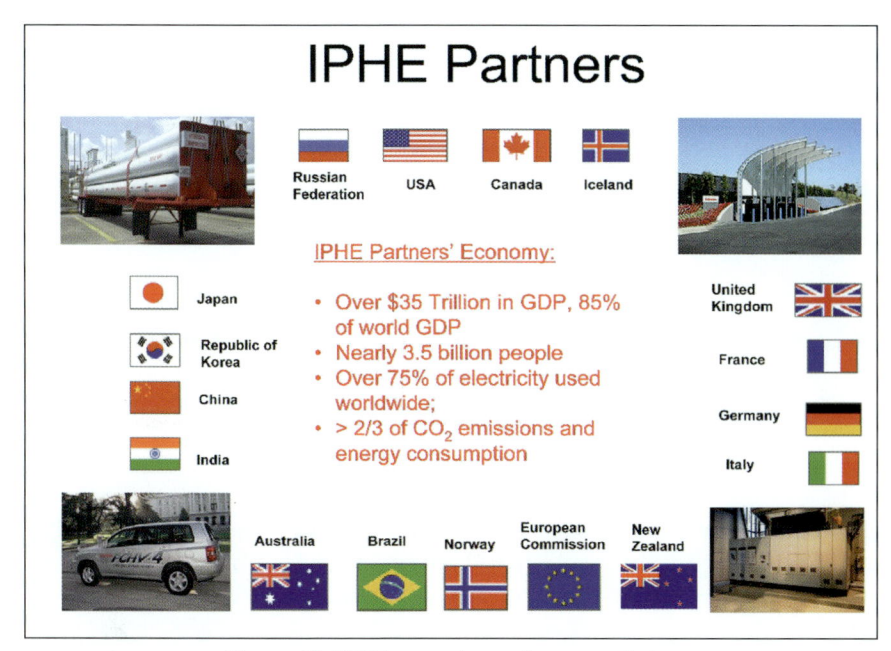

Figure 62. IPHE countries and some statistics.

demonstration and commercial utilisation activities related to hydrogen and fuel cell technologies. It also provides a forum for advancing policies, and common codes and standards that can accelerate the cost-effective transition to a global hydrogen energy economy to enhance energy security and environmental protection".

The envisaged functions of the IPHE was originally to Identify and promote potential areas of bilateral and multilateral collaboration on hydrogen and fuel cell technologies; to analyse and recommend priorities for research, development, demonstration, and commercial utilisation of hydrogen technologies and equipment; to Analyze and develop policy recommendations on technical guidance, including common codes, standards and regulations, to advance hydrogen and fuel cell technology development, demonstration and commercial use; to Foster implementation of large-scale, long term public-private cooperation to advance hydrogen and fuel cell technology and infrastructure research, development, demonstration and commercial use, in accordance with Partners' priorities; to Coordinate and leverage resources to advance bilateral and multilateral cooperation in hydrogen and fuel cell technology research, development, demonstration and commercial utilisation; and last but not least to address emerging technical, financial, legal, market, socioeconomic, environmental, and policy issues and opportunities related to hydrogen and fuel cell technology that are not currently being addressed elsewhere.

The basic structure of the IPHE was based on two main committees: A Steering Committee and an Implementation and Liaison Committee. At the Washington meeting the United States was elected as head of the Steering Committee, originally chaired by David Garman, Undersecretary of Energy, and Germany and Iceland elected as CoChairs

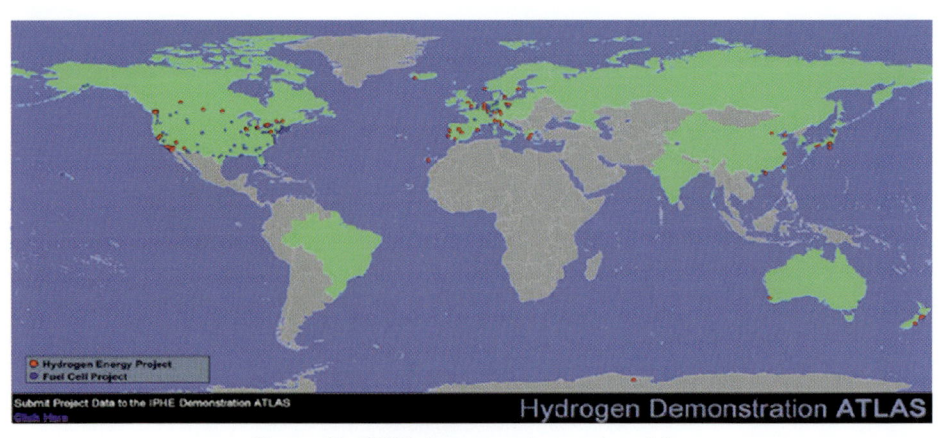

Figure 63. IPHE demonstration project Atlas.

of the Implementation and Liaison Committee. The German CoChair was Hanns-Joachim Neef. The Secretariat of IPHE Has been situated in Washington DC where the vicinity of the Department of Energy headquarters has proven a great asset to the partnership.

During the first two years of the IPHE work a number of products and new services were delivered to the partners. A continuously updated Website has been kept as a living voice of IPHE, www.iphe.net. The website is a harbor for the so called World Demonstration Atlas, the IPHE Brochure and Hydrogen Fact Sheets and Hydrogen Vision and Roadmap (with members). IPHE has devoted a lot of work to setting the scene for future work by the production of the IPHE "Scoping Papers" (5) and IPHE Project and Events Guidelines. A particular attention has been given to stakeholders, international workshops and common projects run by two or more IPHE member countries. The interested reader is referred to the webpage for further deepening of the knowledge about this global network of ministerially based hydrogen cooperation. The Secretariat in Washington was until 2007 headed by Mr. Mike Mills who tirelessly oversees the day to day operation of the IPHE.

In late 2007 Canada took over the chairmanship of the Steering Committee of IPHE and France together with UK the role of chairing the Implementation and Liaison Committee. Sara Filbee and Graham Campbell chair the SC, John Laughead and Paul Lucchese are new ILC co-chairs.

PART V
AROUND THE WORLD IN EIGHTEEN LEAPS

To give the reader an insight into the development of hydrogen technology and hydrogen energy economy at national level, I intend to go through a number of countries, mainly those in the International Partnership for the Hydrogen Economy, IPHE. We start with the impressive work done in the southern hemisphere and then go in nearly alphabetical order from country to country to, in a very short manner, with some of the main national issues and thus cover most of the world.

Australia, New Zealand and Antarctica and the concern for the planet

I have always admired the hydrogen energy interest in Australia and New Zealand. In May of 2003 I took part in the first Australian hydrogen energy conference in the beautiful setting of Broome north of Perth. The active participation of Australian politicians in this conference was a stark reminder for me about the seriousness of the national urge for action.

Australia boasts tremendous endowments of oil, coal and natural gas or the equivalent of four billion barrels of oil, 2.52 trillion cubic metres of natural gas and 79 billion tons of coal reserves as of 2005. Thus, Australians do not need to fear running out of fossil energy stock, in spite of being great exporters of energy to their energy hungry neighbour in the north, China.

New Zealand, on the other hand, has 30 billion cubic metres of natural gas and 8.6 billion tons of coal reserves. In short, those two countries are well endowed with fossil fuel reserves – yet are very interested in developing a new fuel base and to reduce their reliance on the use of such fuels.

As early as 1998, the City of Liverpool in the Greater West of Sydney, became the first region in Australia to acknowledge hydrogen as a clean fuel which would greatly benefit the environment. The declaration owes its creation to Stephen Zorbas who also initiated the National Hydrogen Institute of Australia. In July 2005, the *Australian Hydrogen Activity* report was released by the Ministry for Industry, Tourism and Resources, building on the *2003 National Hydrogen Study*. The report describes Australian research into hydrogen technologies and related issues by covering the full supply chain from the production of hydrogen, through storage and distribution to final use. It is reported that more than 100 projects are being undertaken in Australia, involving at least 36 different organisations. The report will become a valuable resource for those wishing to identify international collaboration partners. It will be interesting to watch the policies of the new Prime Minister, Kevein Rudd, elected late in 2007.

Significant hydrogen demonstration projects are also a part of the activity in progress in Australia. In collaboration with the Clean Urban Transport for Europe (CUTE) project, three hydrogen fuel cell buses were tested on normal service routes of Perth, Western Australia under the Sustainable Transport Energy Perth (STEP) project which started in 2004. The STEP was sponsored by the Australian Government, State Government of Western Australia, DaimlerChrysler and BP and the progress evaluated by Murdoch University in Perth. After completing the initial two year programme, operations were extended for another year to September 2007 with additional 1.75 million Australian dollars (AUD).

Meanwhile, at Mawson base in Antarctica, Australia has tested wind energy-based hydrogen in a project linked to the Australian Government and the University of Tasmania since the installation of a powerful wind turbine system in 2002. The combination of two wind turbines and hydrogen is expected to eliminate the need to transport diesel to the remote location and reduce Australia's greenhouse gas emissions in Antarctica.

In addition, the important Commonwealth Scientific and Industrial Research Organisation (CSIRO) established the Energy Transformed National Research Flagship in October 2003 to respond to environmental and efficiency challenges facing the energy sector. The CSIRO is undertaking key research, in collaboration with other national and international research institutes, in the use of solar energy to produce hydrogen from methane-containing gases. In March 2006, the National Solar Energy Centre (NSEC) was opened by the Minister for the Environment and Heritage. The NSEC, based at CSIRO's Energy Centre in Newcastle, plays a key role in CSIRO's ongoing research and will be home to the largest high-concentration solar array in the Southern Hemisphere.

Australian universities have made an impressive contribution to hydrogen research and development, notably University of Queensland, Brisbane, Curtin University of Technology, Australian National University, Griffith University in Brisbane, Monash University in Melbourne, University of New South Wales in Newcastle, University of Tasmania in Hobart and more. An important textbook on fuel cell systems written by Andrew Dicks of Queensland University and James Larminie is in use all over the world.

New Zealand has a long tradition of providing much of its energy from local, renewable sources. In 2005, 29 per cent of total primary energy supply (TPES) and 64 per cent of electricity in New Zealand came from renewable sources. The Sustainable Development Programme of Action released by the New Zealand government in January 2003, focused on energy issues.

The Ministry of Economic Development is responsible for shaping New Zealand's energy technology development policy and the government, through the Foundation for Research Science and Technology (FRST) – the key agency with primary responsibility for investing in publicly-funded research has invested in hydrogen energy pro-

Figure 64. A scene from Australian Mawson Antarctic Station in Anctarctica.

grammes such as Hydrogen Energy for New Zealand (2002-8), Renewable Distributed Energy (2002-8), Renewable Fuels for New Zealand (2002-4) and New Materials for the Hydrogen Economy (2004-6), covering scenario modelling, production, storage and utilisation. The largest of these programmes, Hydrogen Energy for the Future of New Zealand, is a 1.3 million New Zealand dollars (NZD) per annum programme with two main themes. The first is to generate economically viable fuel cell grade hydrogen from New Zealand's abundant reserves of low rank coal; the second to generate hydrogen from distributed renewable energy resources via electrolysis, with particular emphasis on wind-based generation. The programme is overseen by a Governance Panel of senior executives from the energy industry.

In December 2006, the government released the draft *New Zealand Energy Strategy to 2050* that would put the country firmly on the path towards a sustainable, low emissions energy future. In the draft, hydrogen was recognised as an area that could make a significant impact on achieving sustainable energy objectives and therefore it was recommended that a national hydrogen capability be maintained, niche applications pursued and developments in hydrogen and fuel cells actively monitored with a focus on international collaboration.

Most of the hydrogen research is undertaken by Crown Research Institutes (CRI), universities and research associations. In particular, Industrial Research Limited (IRL), a Crown Research Institute, has a high level of involvement in new and renewable energy research. The company, CRL Energy and Massey University are also involved in hydrogen energy research. CRL Energy is responsible for the hydrogen production from low rank coal theme of the Hydrogen Energy for the Future of New Zealand programme and IRL for the distributed renewable theme. In this way the country designs a future involving both a modernised coal scenario and the renewable energy pathway.

As regards demonstration projects, Industrial Research Limited (IRL) is running a pilot project at Totara Valley, a small farming community in the lower North Island. The project includes a hill-top wind turbine generator and electrolyser to produce hydrogen gas, connected to a fuel-cell and hydrogen burner for water heating at the farmhouse in the valley below. The wind source is used to power the electrolyser to produce hydrogen gas and the project uses approximately two km of pipeline which provides five MWh of hydrogen energy storage. New Zealand has also completed a PEM residential fuel cell demonstration project in Christchurch in cooperation with US Department of Defence and is involved in a similar SOFC demonstration project in the Gracefield Research Centre in Wellington. Because of the relatively low population density and sparse electricity networks in New Zealand, the development of interactive microgrids based on micro CHP fuel cell systems is of considerable interest.

Brazil setting the pace for a continent

When in Sao Paulo in 2007, I was reminded of the enormous pace of urban development in South America. The city of some twenty million inhabitants has over 125,000 streets and there are over one thousand new streets built every year. Streets, cars and people all call for new thinking in energy and transport.

Brazil has showed interest in hydrogen research and development since 1980. Around the turn of the millennium, Brazil produced and consumed over 425,000 tons of hydrogen in the form of fertilizers in oil refineries and the steel industry. Beside this, the Brazilian sugar cane industry is one of the largest sources of potential bio energy in the world, with over 318 million tons of sugar cane, half of which goes to the production of bio ethanol. Hydrogen is expected to complement renewable energy source in Brazil, which accounted for about 44 per cent of the total primary energy supply (TPES) in 2005. A blooming bioethanol industry in Brazil makes it hard for hydrogen to compete there.

Ministry of Mines and Energy (MME) and Ministry of Science and Technology (MCT) have exerted effective leadership in building domestic hydrogen research networks. The MME has operated the Hydrogen Economy Programme focusing on market assessment and commercialisation of hydrogen technologies and providing R&D demands. The Ministry also published the *Brazilian Hydrogen Roadmap* in March 2005 to plan and develop actions that may lead to hydrogen use by 2020 as a complement to renewable energy mix in Brazil. Meanwhile, the MCT is actively funding fuel cell and hydrogen projects and was responsible for the founding of the Brazilian Fuel Cell Programme (ProH2) in August 2002 which will last till 2012. In the programme, the MCT leads science and technology activities under five networks.

The Brazilian construction of a roadmap allowed the elaboration of diagnosis of the hydrogen sector and helped identify four priority programmes: ethanol reforming, water electrolysis, natural gas reforming as well as alternative process. First, ethanol reforming to hydrogen and its direct use on fuel cell were presented as technologies that must be developed with highest priority considering the expertise related to sugarcane processing. Second, electrolysis of water was expected to become a hydrogen source since more than 70 per cent of generation capacity comes from hydroelectricity. Third, natural gas reforming systems were expected to be optimised in order to create initial hydrogen market since natural gas would be the most widely used source for hydrogen production in the first 15 years. Finally, as alternative processes, biomass gasification, biological and photoelectrochemical processes were identified as promising options for hydrogen production in Brazil.

There are a number of players in the Brazilian hydrogen energy economy. LACTEC (Institute of Technology and Development) has worked with a number of fuel cells in the 200kW range since around the beginning of the 21st century. Companies like ElectroCell and Unitech have developed PEM prototypes up to 50kW. Petrobras, the Brazilian energy company, is involved in all steps of the hydrogen chain such as hydrogen hybrid fuel cell bus demonstration, R&D on new conductive membranes and solid oxide fuel cells technologies. In addition, ten universities are participating in hydrogen research and, in particular, Federal University of Rio de Janeiro (UFRJ) and the State University of Campinas (UNICAMP) are leading such efforts. Brazil has also has started the development of codes and standards for hydrogen and fuel cells led by ABNT (Brazilian Association for Technical Standards).

In November 2006, the country's first hydrogen fuel cell bus project was officially launched by the MME, in partnership with the United Nations Development Programme (UNDP), the Global Environment Facility (GEF) and the Federal Foundation for the Brazilian Research and Development (FINEP). The 16-million USD project consists of the purchase, operation and maintenance of up to five vehicles and a station for hydrogen production based on water electrolysis and fuel supply for the buses, expected to start running in an experimental phase in Sao Paolo in 2007.

Figure 65. A sugarcane and bio-ethanol factory in Brazil.

Canada, the impressive hub of hydrogen technology

Canada is a country of extremely varied energy resources spanning from hydroelectricity through nuclear energy to fossil fuels. The country is a net exporter of energy and has seen increases in production and export far beyond domestic use and Canada is one of the most important sources for U.S. energy imports.

In many ways, Canada fosters one of the cradles of hydrogen technology in the world. Canadian companies have risen to world leading position which is heralded by such names as Ballard Power Systems, Hydrogenics Corporation, Stuart Energy and Angstrom Power. Important clusters have been built in Canada around hydrogen technology, with more than a hundred companies contributing some way or another.

The Canadian government is a strong supporter of hydrogen industry, providing a direction for commercialisation and assisting industry in making capital investment in r&d. In particular, Industry Canada, Natural Resources Canada and National Research Council are noteworthy. Industry Canada published the *Fuel Cell Commercialisation Roadmap* in March 2003 developed through the participation and assistance of industry and academia. The roadmap represents a critical step in identifying the commercialisation challenges and in selecting the strategies and actions that will allow Canadian stakeholders to successfully meet the challenges. The Industry Canada also developed a national strategy, *Towards a National Hydrogen and Fuel Cell Strategy for Canada* in 2006 which presented a long term vision for Canada's participation in the hydrogen energy economy together with a phased action plan. The Natural Resources Canada (NRCan) promotes the use, development and production of alternative transportation fuels such as hydrogen fuel cells and supports a hydrogen refuelling demonstration initiative called Canadian Transportation Fuel Cell Alliance (CTFCA). Moreover, the NRCan assists with the development, demonstration and evaluation of fuel cells, hydrogen fuelling systems and hydrogen storage technologies. The National Research Council (NRC) is also a major player, in that the NRC runs the Fuel Cell and Hydrogen Programme, focusing on three strategic areas: Polymer Electrolyte Membrane Fuel Cell; Solid Oxide Fuel Cells; and Hydrogen Generation and Infrastructure. Beside these entities, Technology Partnership Canada, Environment Canada, Transport Canada as well as Social Sciences and Humanities Research Council are supporting various efforts on the subject.

The determination of Canadian industry is globally outstanding. The industry is leading R&D advancements, investing over one billion Canadian dollars (CAD) in R&D. The industry includes over 80 companies and organisations, employing more than 2,200 people across the supply chain. Total revenue of the industry in 2004 reached 133 million CAD and total R&D expenditure in the same year was 237 million CAD or over 100 thousand dollars per employee. It is worth mentioning the achievements of major Canadian companies. Angstrom Power demonstrated portable electronic devices run on hydrogen and fuel cells. Ballard Power Systems showed good progress in PEM fuel cell development. A significant milestone is 50 consecutive freeze starts at minus 20 °C operational temperature; fuel cell stack durability of over 2,200 hours aiming for 5,000 hours in 2010; cost reduction in the stacks by reducing platinum catalyst by 30 per cent. Hydrogenics Corporation markets a 10kW power module which has achieved technology milestones for critical backup power and fuel cell hybrids as battery replacements for forklift trucks.

Thus, it is not surprising that there are a number of significant demonstration projects underway in Canada and here are some examples.

First, Hydrogen Highway project aims to enable and advance the use of Canadian hydrogen and fuel cell technologies. The network of Hydrogen Highway develops not only refuelling stations but also mobile, stationary, portable and micro-fuel cell applications throughout British Columbia's southwest corner. As part of this project, four hydrogen fuelling stations are operated in Victoria, Surrey, North Vancouver and Vancouver, and more stations are to be built in the area. BC Transit also developed a transit business plan to bring 20 hydrogen fuel cell powered buses to Whistler for the largest pilot project of its kind, starting in 2009 aimed at the 2010 Olympic Games and Paralympics Winter Games.

Second, the Hydrogen Village project seeks to raise awareness and break down barriers to markets for hydrogen, fuel cell and other relevant technologies within the Greater Toronto Area. The concept of the project was proposed in 2003 and it is sponsored by the Government of Canada's Hydrogen Early Adopter's Programme, University of Toronto at Mississauga (UTM), and Fuel Cell Technologies Ltd. Through this project, three refuelling stations were installed in the area and Hydrogenics Corporation demonstrated two hydrogen and fuel cell powered forklifts in regular service. The UTM opened Canada's first multiple solid oxide fuel cell facility in April 2006 supplying clean and environmentally friendly heat and power for 12 student townhouse units.

Figure 66. Geoffrey Ballard. Source: Ballard.

Third, the Vancouver Fuel Cell Vehicle Programme is to promote public awareness and demonstrate zero-emission transportation and started with the arrival of five Ford Focus fuel cell vehicles in Vancouver in April 2005, four of those were operated in Vancouver and one in Victoria. These hybrid-electric vehicles use the Ballard fuel cell engines and Dynetek storage for compressed hydrogen. Three fuelling stations for those vehicles are located in Surrey, Vancouver and Victoria. Finally, the Prince Edward Island Wind - Hydrogen Village Demonstration Project will demonstrate, in real-life and in real-time, how wind energy and hydrogen technologies can work together to offer clean and sustainable energy solutions across a wide range of applications, including the installation of a hydrogen energy station, a H-storage depot, and a wind-hydrogen and wind-diesel integrated control system. This project started in April 2005 and involves 15 organisations. Technology Partnerships Canada will invest 5.1 million CAD, contributing about half of the total project costs of 10.3 million CAD.

Canadian authorities have played their cards well and industry has also fulfilled its part successfully. The Canadian hydrogen and fuel cell businesses were spending about 50-60 million CAD from public funding in 2006 and some 200-240 millions from private sources, totaling up to about 300 million CAD. Notice the larger proportion of private capital! With due consideration of its enthusiasm, Canada is likely to maintain its key position in the hydrogen technology market for the foreseeable future.

China could leapfrog and lead the way

China with its large population and thriving industrial development coupled with concern for the environment is becoming an ever increasingly important player on the hydrogen world scene. The scenario is vast in size and expansion. To name a few indicative statistics, we note that the annually added office space in Shanghai alone amounts to eleven times the World Trade Centre in New York. Every day over one thousand new cars begin their service life on the streets of Beijing and the middle class in China is gaining enormous purchasing power. The Chinese are planning the future and hydrogen is an exciting option in this largest nation of the world.

The total production of hydrogen in China around the turn of the century was eight million tons. 97 per cent of this hydrogen was made from fossil fuels and 3 per cent by water electrolysis. This production aside, China annually recovers some 0.3 million tons of hydrogen off-gas from ammonia plants and some 0.1 million tons from chlor-alkali plants. Professor Zong-Qiang Mao at INET institute, Tsinghua University, estimates that some 4 billion RMB (half a billion USD), potential construction cost for hydrogen production infrastructure, could be saved by using this hydrogen off-gas. Since over 69 per cent of Chinese primary energy supply came from coal, gasification of coal is a very important method of producing hydrogen. Some four million tons of hydrogen are produced by coal gasification annually. China acknowledges the pollution problem resulting from coal use and is striving to increase the role of natural gas and oil as feedstock for hydrogen production. The Chinese authorities envision that more hydrogen will be produced from renewables such as solar hydrogen and wind energy produced on-site as well as from agricultural biomass gasification and the use of hydroelectric energy for electrolysis. China also recognises the link between natural gas and hydrogen and regards it as an important factor in future energy policies.

The Ministry of Science and Technology (MOST) is in charge of government R&D investment and allocates funds to state-owned institutes, universities and spin-off institutes. MOST initiated the National High-tech R&D Programme (863 Programme) in March 1986 and the National Basic Research Programme (973 Programme) in March 1997. Both of the programmes include the themes on energy and new materials, and MOST has launched projects on hydrogen production, storage, refuelling station and fuel cells in the 10th Five-year Plan period (2001-2005) and the 11th Five-year Plan period (2006-2010).

As regards hydrogen fuel cell vehicle development, China has paid close attention to PEM fuel cell R&D since 1990s, relying on researchers from Dalian Institute of Chemistry and Physics, Beijing Fu Yuan Pioneer New Energy Material Company, Shanghai Sun Li High Technology Company and Beijing Lu Neng Power Sources Company.

The first Chinese PEMFC vehicle put on the road in 1999 was a cart developed by Tsinghua University using 5kW fuel cell stacks. Since then, several prototypes of fuel cell vehicles have been made. In particular, three demonstration projects for hydrogen and fuel cell vehicles are prominent in China, namely, the Fuel Cell Bus project, named GEF/UNDP, a domestic fuel cell development project and Beijing Hydrogen Park. The project sponsored by international funding was incepted officially in Beijing in March 2003 to promote sustainable transportation in the developing countries. Three fuel cell buses purchased from DaimlerChrysler began demonstration operation in Beijing in June 2006. Aside from this, the MOST ministry supported domestic fuel cell vehicle development projects, providing 106 million USD during the 10th Plan period (2001-2005). Going through three phases of development between 2002 and 2005, a total of five fuel cell buses and ten cars were developed by Tsinghua and Tongji University.

Figure 67. Dalian Institute in China, one of the country's major centres for hydrogen research and development.

The Beijing Hydrogen Park is China's first demonstration project for new energy vehicles. The park, comprising a research and development centre, a hydrogen refuelling station, a fuel-cell vehicle garage and a maintenance workshop, will provide critical experience in the infrastructure needed to operate fuel-cell vehicles. It will oversee several international trial programmes and also fuel the hydrogen vehicle fleet for the 2008 Beijing Olympic Games. Indeed, the Olympics are shaping up as a milestone test for Chinese hydrogen and fuel cell culture. At the same day when the Beijing Hydrogen Park opened in November 2006, China's first hydrogen refuelling station went into operation. The station will be operated by world energy giant BP, covering an area of 4,000 square meters. Thus, China entered the international fuelling station scene and expects 10 refuelling stations by 2010 and 4000 by 2020.

The European Commission and the impressive policy work

Good old Europe was the foster place of the hydrogen pioneers. The affection of Europe´s past is reflected in the policies of the European Commission (EC) towards hydrogen. This emphasis is clearly emphasized in the fact that within the Framework Programmes (FP) for research and development, the Commission has increased the funding for hydrogen research and technical development by approximately doubling the budget every four years. The FP 6 (2002-2006) required 300 million Euros for hydrogen; the FP 5 (1998-2002) demanded 145 million Euros; and the FP 4 (1994-1998) spent about 58 million Euros – doubling again the contribution to the FP 3 (1990-1994) and so on. The expected breakdown of the 300 million Euros goes to various branches of hydrogen R&D, the largest proportion to validation and demonstration, a little less to transport applications, including FC vehicles and the third largest contribution to the production of hydrogen.

When Romano Prodi of Italy was the President of the European Commission, he is

Figure 68. Research Commissioner Philippe Busquin at a fuel cell bus event in Europe. Source EU Research.

quoted to have said in relation to white papers on energy from 2000: "For us, reducing fossil fuel dependency is a priority." Prodi was preoccupied with the element of European competitiveness, and he said, when delivering his message on a dramatic increase in spending on renewable energies, that central focus would be hydrogen fuel cells, a "field where the union has lagged the United States and Japan in publicly financed research."

Following this forceful intervention, Europe has seen a flourishing research and development on hydrogen and fuel cells. In the turmoil of victory, President Prodi furthermore said, that Europe was poised to leap ahead of its rivals in its overall energy strategy. He said "Neither the United States nor Japan is clear on its goals, and without clear goals, there is little progress."

What was particularly positive about this intervention is that it made Europe unusually straight about its goals and created a spirit of progress not often seen in the continent. In the action of the Commission, there was a clear potential for reducing dependence on oil import as well as for greenhouse gas reduction beyond Kyoto targets, especially in case of use of renewables as primary energy sources. The firm policy was expected to contribute to 20 per cent substitution of diesel and gasoline by alternative fuels in 2020, as articulated in the White Paper on EU Transport Policy, while still keeping the potential for promoting industrial competitiveness.

Following recommendation by the High Level Group (HLG) on Hydrogen and Fuel Cells launched in October 2002 and which published the *Hydrogen Energy and Fuel Cells - A vision of our future* in 2003, the Commission established the European Hydrogen and Fuel Cells Technology Platform (HFP). The vision paper was followed by *Strategic Research Agenda* and *Deployment Strategy*, both of which were presented and endorsed at the 2nd General Assembly of the Platform in March 2005. Certainly, a very important movement with few equals in the history of hydrogen energy.

The latest framework programme, FP 7 (2006-2013), is expected to include a billion Euros over seven years, as proposed of the Commission in April 2005. One of the new aspects of the FP7 is called European Joint Technology Initiative (JTI). This time, the programme will be led by a public private partnership with basic funding from the Framework Programme. The remaining funding will come from industry, national, regional and local programmes, structural funds, etc. What the Commission is anticipating is that this new structure will allow a more efficient and flexible implementation of the *Strategic Research Agenda and the Deployment Strategy*.

We can look at the *Strategic Research Agenda* in view of what we have gone through in previous chapters. The paper calls for the highly focused ten year r&d programme with the following aims: to reduce fuel cell costs down to a tenth or a hundredth part (depending on applications); to enhance performance and durability of fuel cell systems by a factor of two or more and to reduce costs of hydrogen production and delivery by a factor of three or more. The analysis further points out the need to combine private and public investment at European scale – including the EU countries, individual member states and regions – and really to double present effort. Furthermore, the report stresses the need to develop policy frameworks and financial schemes and to bridge the gap between research and development and commercialisation.

The analysis concludes that the aim would be to establish early markets – including portable applications – which could be operating by 2010, with stationary applications achieving commercialisation by 2015 and mass transport applications by around 2020.

The Commission views the Joint Technology Initiative (JTI) as a new way to realising public-private partnership in the continent. In the 10 year programme, it is expected that the JTI will manage funds exceeding 6.7 billion Euros. The idea is to let industry lead this partnership in a structure which will ensure that the jointly defined research programmes will better reflect the needs and expectations of the European industry. The research community in Europe will also be represented among the founders of JTI. In this way, the EC hopes to travel the road with maximum acceleration towards success. The JTI was finally established in he autumn of 2007.

France – The post-Carnot era

A hydrogen energy revolution needs France who has a strong willingness to develop international cooperation and supports actively the afore mentioned European Hydrogen and Fuel Cells Technology Platform. The importance of hydrogen R&D is continuously increasing in France, especially because the French government decided to reduce greenhouse gas emissions by 75 per cent in 2050. In 2004, the French authorities commissioned a review of the R&D priorities in France, and the report of a working group, published in May 2005, included a suggestion that France should aim to be a key partner in hydrogen and fuel cells, photovoltaics and others. The conclusions of the report immediately led to the establishment of the New Energy Technology R&D Programme which has five priorities: hydrogen and fuel cells; CO_2 capture and storage; photovoltaics; energy efficiency and bioenergy. Thus, the National Action Plan for hydrogen and fuel cells (PAN-H) was launched in 2005. The PAN-H comprises three stands of industry/academia research. To begin with, the plan aims to develop the PEM fuel cell technologies necessary for automotive applications, including on board hydrogen storage. At the same time, the plan will investigate pipeline materials, model their behaviour and develop economic models for different development scenarios. The plan aims to investigate high temperature production of hydrogen by electrolysis or thermo-chemical cycles. 200 million Euros from public funds will be allocated to the plan between 2005 and 2010.

A world known organisation in France is the Association Lorraine for the Promotion of Hydrogen and its Applications, (ALPHEA), a strategic research centre for fuel cells and hydrogen. Founded in 1975, the centre under the leadership of Rene Sachs, supports the local industry with scientific analysis. The centre is influential in policy creation in France and beyond. The French Hydrogen Association AFH2 was established in 1998 and is currently led by Thierry Alleau.

The establishment of the National Research Agency (NRA) and the Agency for Industrial Innovation (AII) in 2005 is an important milestone for hydrogen and fuel cell research policy in France. The NRA, created by Ministry of Research, focuses on basic and applied research and has a major impact on developing public private partnerships. The AII, formed by Ministry of Industry, is engaged in industrial research programmes

Figure 69. A team from CEA France demonstrating a concept hydrogen car.

involving R&D and has been discussing large scale demonstration projects with industry.

Electricité de France (EDF) and Gaz de France (GDF), two leading utilities companies in France, have demonstrated fuel cells of various kinds over a long time using natural gas, methanol and biomass as primary fuel. The companies have jointly operated a demonstration project in the Paris suburb of Chelles since 2000, where they test 10-200kW phosphoric acid fuel cells (PAFC) system, supplying power to about 200 homes. The EDF is also working with a European consortium to test a fuel cell in the heating plant of the Treptow district in Berlin and is participating in the several EU projects such as AUTOBRANE and Real-SOFC. Meanwhile, the GDF initiated the EPACOP project in 2002 aimed at testing five residential PEM fuel cells powered by natural gas for two years in real life conditions. With financial support of the Agency for Environment and Energy Management (ADEME), those fuel cells were tested on five different sites in France: two fuel cells in Dunkerque, the others in Sophia-Antipolis, Nancy and Limoges.

As regards new products, PSA Peugeot Citroën in collaboration with the French Atomic Energy Commission (CEA) developed the new world-class fuel cell stacks for automotive applications in 2006 through the GENEPAC (Fuel Cell Electricity Generator) project, featuring compact and efficient fuel cell stack with a rating of up to 80 kW and cold start at minus 20 degree centigrade. PSA Peugeot Citroën aims to be a global technology leader in automotive range extender solutions before 2009.

Another area to watch in France is the research sphere of CEA and Air Liquide, the large international company. These giants are, on the one hand, experts in nuclear energy technology and, on the other, in hydrogen gas/liquid production and infrastructure respectively. In fact, France is a stronghold of knowledge on the utilisation of nuclear energy for hydrogen production and a number of ways are being examined for high-temperature electrolysis using nuclear energy. This work has been extended to the inclusion of other heat energy sources like geothermal energy. Moreover, it cannot be overlooked that some companies such as Air Liquide, Axane and Helion Fuel Cells are already supplying fuel cells. Finally, CEA and IFP (French Petroleum Institute) with LBST (L-B-Systemtechnik GmbH) in Germany have created E3database, an informational database to promote a well-to- wheel economic evaluation of hydrogen chains in key energy sectors. This database is expected to function as a decision-aiding tool for the evaluation of a wide range of hydrogen fuel chains.

Germany, masters of endurance

Germany enjoys a special position regarding hydrogen and fuel cell technologies. German companies have for a long time been among the most active hydrogen development platforms in the world and have been able to carry this out, independent of federal government support. R&D on hydrogen and fuel cell technologies has been one of the key areas of the *Programmes on Energy Research and Technologies* of the Federal Government for a long time. Additionally, in Germany, we find very active regional activities in most of the Federal States such as Baden-Württemberg, Bavaria, Hamburg, Northrhine-Westphalia and others enjoying support from local or federal governments.

Until the middle of 1980s, Germany focused on the thermo-chemical production of hydrogen from nuclear energy using the high temperature reactors. Another technology under investigation was the high temperature electrolysis (HOT ELLY) pioneered by a group in Southern Germany led by Wolfgang Dönitz and his colleagues. The general results of the HOT ELLY project were not seen to justify further work at that time, mainly due to enormous difficulties associated with materials degradation caused by the hot environment and what was commonly described as "a plumber's nightmare".

Another German hydrogen project was the impressive HYSOLAR, a solar hydrogen project initiated at the DLR German Aerospace Centre in Stuttgart in 1986 with a 10kW test and research centre. The project, led by Andreas Brinner, involved universities in Jeddah and Riyadh in Saudi Arabia with hydrogen mostly produced by electrolysis.

This was the start of a period that signified nothing less than the first modern heydays of German hydrogen research, including investigations of large hydrogen production in Canada with transport by huge tankers to Europe and the appearance of important engineering text books by for example Carl-Jochen Winter and his colleagues.

In 1988, an Ad-hoc Working Group, commissioned by the Federal Ministry of Education and Research, advised that policy should identify barriers to a hydrogen energy economy and initiate measures to overcome them. In the wake of this recommendation, intensive R&D projects took place between 1988 and 1995 supported with public funds of about five to seven million Euros per year. Work concentrated on the development of specific hydrogen technologies across production, electrolysis, storage and applications, as well as on larger projects to demonstrate the whole supply and usage chain. HYSOLAR, in cooperation with Saudi-Arabia, and BAYSOLAR in Bavaria were the two major projects. Other projects like PHOEBUS at the Research Centre Jülich, demonstrating solar hydrogen, were funded by the Federal States.

In a review made in 1995, the conclusion was reached that the major precondition for a hydrogen energy economy is the production of CO_2-free hydrogen. However, due to the high price, "clean" hydrogen from renewables, especially from solar, was not regarded being competitive for a long period of time. Direct use of electricity from renewables would lead to a more economical path instead of using "clean" power for hydrogen production. Seen in this way, the use of hydrogen as an energy storage option makes sense only if the storage capacity of the grid will not be sufficient to use the electricity produced from renewables. No new solar hydrogen demonstrations were authorised at this period and attention was given to fuel cell development with the expectation that the economic relevance of the (solar) hydrogen energy economy would take 30-50 years. This, rather pragmatic view, has not necessarily been shared by all and has created a certain division in opinion which has been felt as different views in different ministries in Germany.

Fuel cell R&D in Germany started early with the first Federal programme on energy research in 1974 and was intensified after the review in 1995. In addition, a "Programme on Investment into the Future (ZIP)" began in 2001 to accelerate the development and deployment of key technologies, with a special emphasis on field tests for fuel cells for residential and transportation applications.

The option of generating CO_2 – free hydrogen from fossil energy sources, especially coal, entered the Federal Programme on Energy Research and Technologies. The capture and storage of CO_2 is part of the German COORETEC (CO_2 Reduction-Technologies) research concept developed during 2002/3. In June 2003, this concept was published by the Federal Ministry of Economics and Technology. As a related pathway, hydrogen production from biomass is being investigated as part of the bio energy research programme.

A new vision on hydrogen and fuel cell technologies is currently being drawn up on the initiative of the Federal Ministry of Transport (BMVBS), involving the, Federal Ministry of Economics and Technology (BMWi) and the Federal Ministry of Education and Research. The "National Innovation Programme on Hydrogen and Fuel Cell Technologies (NIP)" was published in May 2006 and will be operational in early 2007.

Additional public hydrogen research funding of 500 million Euros over the next ten years will be matched by private investment in R&D and demonstrations. This programme has to be seen in close context with the European Hydrogen and Fuel Cell Technology Platform and its initiatives and projects funded under the 6th and 7th Framework Programme of the European Commission. The federal and state ministries, the German industry and research community are actively participating in the European initiatives on hydrogen and fuel cells.

For national coordination and European and international collaboration, a new agency was established in 2007. It is called "Nationale Organisation für Wasserstoff und Brennstoffzellen (NOW)" and will be acting on behalf of the transport ministry. Together with the Projektträger Jülich (PtJ), currently coordinating R&D projects supported by the economics and transport respectively, the national organisation NOW will play a major role in directing the "National Innovation Programme on Hydrogen and Fuel Cell Technology (NIP)" during the next ten years. NOW will be supported by an advisory council in which federal and state governments, industry and science are represented.

German automobile industry is not only a world leader in its own field but is also deeply involved in the development and market introduction of hydrogen and fuel cell technologies. Daimler and BMW have shown extremely impressive and innovative approaches to the hydrogen energy economy. From the early start, BMW has concentrated on internal combustion engines and liquid hydrogen storage. The Daimler team, on the other hand, started working with methanol fuel cells and later went directly to PEM fuel cells and hydrogen storage onboard, a path that has continued. These two companies have been world leaders in the area of hydrogen and deserve a lot of credit for their initiatives and stamina.

In the *5th Programme on Energy Research and Technologies* of the Federal Government, the emphasis is on development of PEM fuel cell systems as well as high temperature fuel cell systems (MCFC, SOFC) for stationary applications. The emphasis within the SOFC development has been on the tube-form concept as well as on the planar-form concept.

Currently, the annual public funding for fuel cell R&D activities amounts to 8 – 10 million Euros. Projects are cost-shared with the private sector and therefore the figure corresponds to an overall budget of 16 – 20 million Euros annually. In addition to these, Germany is funding fuel cells research by the ZIP programme of the Federal Government. In the ZIP programme, demonstrations play a major role. The ZIP programme 2001-2005 has a volume of about 120 million Euros and is quite far reaching in various types of fuel cells.

When looking ahead up to a decade, Germany anticipates research on development of mature and complete fuel cell systems with targets like costs of 1,000 – 1,500 ¤/kW in stationary applications and 50 ¤/kW in mobile applications, close to a level generally accepted world-wide. Furthermore, a very much improved lifetime of 40,000 hours for stationary and 5,000 hours for mobile fuel cell applications with only minor degradation and the full capacity to compete with other technologies

Germany, being one of the leading hydrogen energy countries of the world also shows some of the most impressive hydrogen energy exhibition and promotion activities. At the annual great industry exhibition, the Hannover Messe, Arno Evers has been organizing a Hydrogen and Fuel Cell exhibition since 1995 with remarkable attendance of the general public of an exhibition which is the most impressive anywhere. Evers started his career with a British oil exploration company in the Persian/Arabic gulf, later joining the German aircraft industry before making a name in the hydrogen energy world. For anyone interested in watching the development of hydrogen energy technology, the Hannover event is a must.

One of the leading hydrogen energy conferences of the world has been held every two years in the old industrial city of Essen, Germany. The presently tidy city has changed considerably since the haydays of coal when the city was one of the power centres of industrious Germany. The hydrogen energy conference banners waiving outside the conference building are a reminder of the new role Essen wants to make in the cleaner energy economy.

Professor Carl-Jochen Winter, the pioneering hydrogen energy advocate in Germany, was the main initiator of the conference in 2002 in an ancient coal mine building.

Hydrogen Infrastructure and Society   157

Figure 70. Carl Jochen Winter with his wife Eva during a visit to Iceland. Source: Icelandic New Energy.

Subsequently there was a conference in 2004 and 2006, and is planned to continue to be held every two years with a grand finale in 2010 when the 18th World Hydrogen Energy Conference will be held in Essen, by then the Culture Capital of Europe.

Winter´s philosophy can be summed up in his own words: "Energy policy is technology politics"! With about 75 per cent dependency on imported energy and placed on the top of a coal mountain, nothing could be more symbolic for the aims of German society than to use the knowledge base of its scientists and engineers to employ the clean energy carrier of hydrogen.

Iceland, perhaps the ideal testing forum

I hope the reader will forgive me for writing a bit more extensive story of the Icelandic hydrogen project which of course is dear to my heart and I feel that so many people deserve our thanks for making this all possible. The island has been a part of the world hydrogen movement since the 1980s with the pioneering work of Bragi Arnason, professor of chemistry at the University of Iceland. In order to better understand the motivations behind the Icelandic scenario, it is of importance to go through a few points of history.

During eleven centuries of habitation, Icelanders have seen various stages of development in energy history. The country, much deprived of fossil energy reserves, relied upon a number of primitive fuels, peat and driftwood for almost a thousand years. Of course, coal, coke and oil were imported to the island in tune with the development of the steam engine and the industrial revolution, and later with the use of combustion engines and automobiles. This all happened amidst the plentiful natural energy resources that Iceland truly possesses.

The intrinsic nature of these natural resources is related to the volcanic nature of Iceland. In fact, the island lies on the so-called "hot spot" on the tectonic boundaries between the Eurasian and the NorthAmerican plates. The plates drift apart below Iceland by something like an inch a year. Magma from the interior of the earth fills the gap with the resulting expansion of Iceland's territory by some square kilometers every century! The access to geothermal energy is made through boreholes. Since around the end of the century, Icelandic municipalities have been close to 100 per cent sustainable as regards space heating by geothermics.

On the other hand, Iceland, due to its geographical position in the pathway of the low-pressure anticyclons entering from the South West, receives a lot of humidity from the Atlantic Ocean. The humidity causes precipitation in Iceland when coupled with the land elevation, and this creates a basis for hydroelectric energy, which today is mostly derived from glacial rivers. One can say that the glaciers are gravitational energy capacitors; they store the potential energy of the precipitation. Some of them even have very high flux of geothermal heat underneath so that the ice mass is more rapidly turned into running water. 80 per cent of the country's electricity stems from hydroelectric harnessing. 20 per cent originates from electric generators connected to steam turbines in geothermal power plants.

Large scale use of hydroelectric energy in Iceland has been linked to the operation of aluminium smelters and other heavy industries, and the government of Iceland argues, quite rightly, that they are based on renewable resources and hence lead to much less gross carbondioxide emissions than similar industries based on coal as a primary resource.

Around the turn of the millennium, Iceland had the highest proportion of renewable energy in its energy portfolio in the world. Then, about 72 per cent of energy originated from renewable sources. The proportional use of various energy types can be seen in Figure 71 where we show the domestic energy portfolio by source. As can be seen from the figure, Iceland depended greatly upon coal in the first decade of the 20th century. This was later displaced by geothermal utilisation. Around 1965, large scale

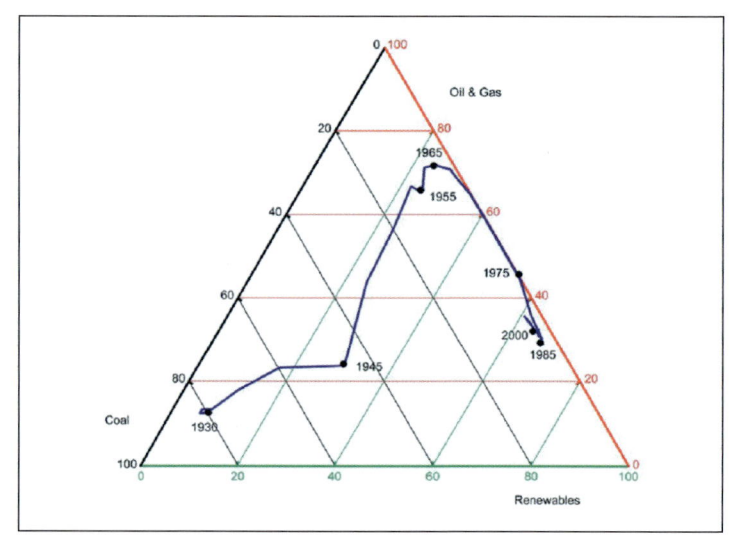

Figure 71 The proportional energy portfolio of Iceland shown as a three phase diagram: Coal/oil-gas/renewables in a timeframe from 1930 up to the end of the century.

Hydrogen Infrastructure and Society 159

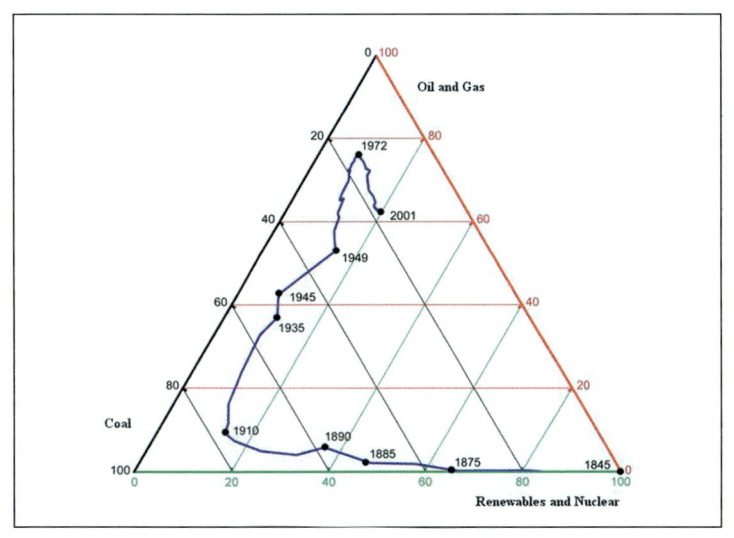

Figure 72 The proportional energy portfolio of USA. We have added nuclear power to keep the three phase diagram representation.

utilisation of hydroelectric energy for metals industries, such as aluminium and later ferrosilicon, increased the use of renewables further. We show a similar figure for the U.S. energy portfolio in Figure 72

Having observed the 72 per cent renewable proportion, may we ask where the remaining 28 per cent come from? The answer is mostly *imported oil and gasoline for transport and fishing*. The transport sector uses roughly about a third of the imported liquid fossil fuels and the fishing industry roughly the same proportion. The remaining third belongs to other industries.

When the Kyoto protocol was discussed and the year of 1990 was set as a baseline, it was clear that Iceland would not enjoy the huge benefits of "cleaning up the fossil based central heating" when fossil fuels were displaced by geothermal energy before 1990. The Government of Iceland was well aware of this and, at the same time, determined to carry on with the use of renewables.

Two major options stood out for replacing fossil fuels in the transport sector. One was the use of electric cars or transport and the other was linked to designed fuels like hydrogen or methanol. The University of Iceland has been involved in studies of all these options. Electric cars have been tested in Iceland with the conclusion that improved battery capacity and charging technology is needed before this becomes an acceptable option.

With the pioneering work of Bragi Arnason, hydrogen was well known as a candidate for an energy carrier. The production of hydrogen had been known since about 1953 when a fertilizer plant was established on the outskirts of Reykjavik. The plant used hydrogen produced by electrolysing water and combined it with atmospheric nitrogen to make a vital ingredient in a fertilizer designed to boost the horticultural quality of the volcanic Icelandic soil and thus enhance the quality of Icelandic grass and hay harvest. For a number of years, the production of hydrogen at the fertilizer plant amounted to 2000 metric tons annually.

In most other economies, the most obvious production route for hydrogen would be natural gas reforming. The obvious difficulties these countries face are that the total pathway of hydrogen is not at all free of carbondioxide emissions. It seemed clear that electrolysis of water by the use of inexpensive electricity from renewables could provide a solution. From a world energy point of view, a hydroelectic/geothermal hydrogen produced in Iceland could at least be a part of a solution in an interim period. While testing hydrogen in Iceland, the world could carry on with improving sequestration techniques, direct solar harnessing efficiency and more demanding options.

Since the start up of the Fertilizer plant in Iceland there has always been an awareness of the potentials of hydrogen as fuel.

And, while dwelling on history, is of interest to digress and reflect back upon the speculations put forth by the futurist writer Jules Verne already in 1874. That year his novel "The Mysterious Island" was published in Paris. Incidentally, 1874 commemorated a millennium of the settlement of Iceland from the time of the first settler, Ingolfur Arnarson in 874AD.

In the novel, an adventurous team of people are discussing the future energy sources and one of the main characters, an engineer, points to water as the energy source of the future! Jules Verne knew and understood the electrolysis of water, discovered around 1800. He knew that water can be split into its elemental components, hydrogen and oxygen by placing two opposite electric poles in the water. The character in the Jules Verne novel defines water as the main energy source of the future society! The Mysterious Island forecast in Jules Verne´s fiction could well have been Iceland as we will see in the following.

There had been a close link between the hydrogen research group at the University of Iceland and a number of German companies and institutions, and there seemed an obvious possibility to team up with the Germans to make further progress in hydrogen technology. Bragi Arnason and myself had been cultivating the link to Germany, and in the early nineties, there was the first Iceland-Germany conference on hydrogen, held in Hamburg under the patronage of the Hamburg Society for the Introduction of Hydrogen Into the Energy Society. Two brothers in Hamburg, the Gretz brothers, were among the main leaders of the society together with Herr Schuess, a shipping magnate and a number of Germans. I also remember a very important link we had with Dr. Kreysa of the German Chemical Industry.

The meeting in Hamburg was attended by a number of our colleagues from Iceland, such as Pall Kr. Palsson who later became the director of the New Business Venture Fund and a main driver for what was to become. The seeds sown in Hamburg were to echo in what later was to be founded.

Shortly after the Hamburg meeting, the Hamburg Society organised with us an exhibition of hydrogen technology at the University cinema in Reykjavik. Links were established with Daimler Benz in Stuttgart, where we enjoyed the help of Ambassador Ingimundur Sigfússon in Bonn.

In 1996, the Minister of Energy and Industry in Iceland created a committee of

Hydrogen Infrastructure and Society

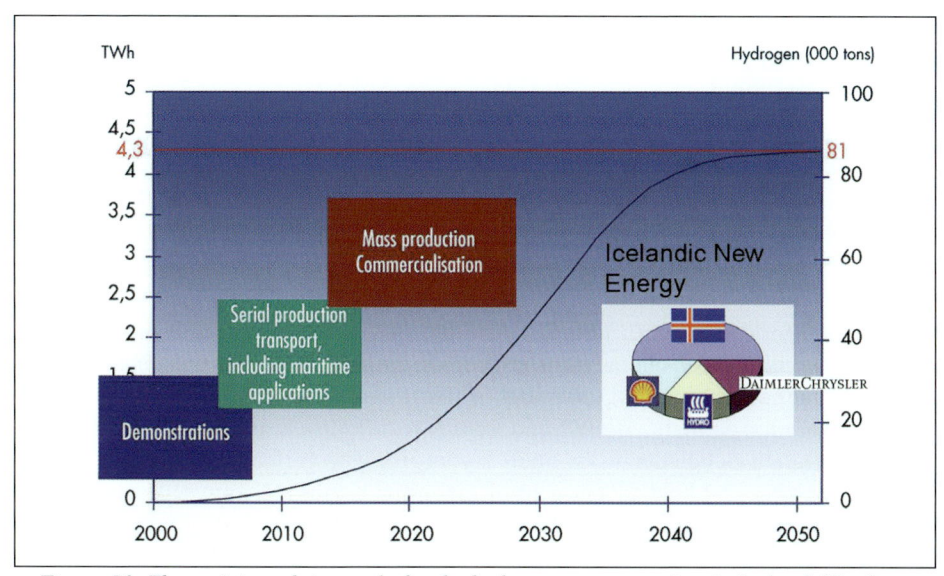

Figure 73. The anticipated timescale for the hydrogen energy society in Iceland. The figure show the amount of hydrogen needed annually as well as the electric energy to produce it. Based on Jon Björn Skulason, Icelandic New Energy.

experts to make suggestions for the domestic production of fuels in Iceland. The aim was how Iceland could use its domestic energy resources to reduce the dependency upon imported fuels. The chairman of the committee was Hjalmar Arnason, a dynamic MP. The committee focused on hydrogen and hydrogen compounds as a candidate carrier. The link with Daimler Benz – Ballard was strengthened in the wake of the committee work and Bragi Arnason and I devised a plan for the introduction of hydrogen into Iceland with the help of Daimler. Bragi and Ambassador Sigfússon visited Daimler in Stuttgart and had discussions with Dr. Panick, by then the head of development of hydrogen technology in Daimler, who argued that cooperation with Iceland would benefit all the partners.

From then on, the basic idea was that the German company Daimler Benz would select Iceland as a site for testing buses which would use electrolytic hydrogen, like the one produced in the Icelandic fertilizer plant, as a fuel. In 1999 the Government declared its intention to aim for a hydrogen energy economy in Iceland which was followed by the creation of a public-private company in Iceland, Icelandic New Energy Ltd (INE). 51 per cent of its shares were owned by VistOrka – a company owned by Icelandic energy companies and institutions and 49 per cent were owned by three international stakeholders; Daimler Benz (later DaimlerChrysler), Norsk Hydro (who were the producers of the electrolysers) and Shell Hydrogen, a company formed by Shell International to pave the way for hydrogen as a commercial energy carrier.

The University of Iceland viewed this as a creation of a spin-off from its hydrogen research. Years of intellectual development in hydrogen energy technology was bearing fruit in an international spin-off company. In the minds of us who had paved the way, it could easily take half a century to create the projected hydrogen energy economy and a number of steps would have to be taken starting with demonstrations which then gradually would evolve into a market driven system. Figure 73 shows the planned timescale of the fifty years plan.

It is estimated that a full scale hydrogen energy economy relating to transport sector and fishing fleet would require about 10 per cent of the total energy potential of Iceland. This potential is at this moment almost eight times greater than exploited energy resources. Powering the transport sector and fishing fleet with hydrogen would require about 80-100 thousand tons of hydrogen annually and over 50 per cent increase in the already harnessed energy.

The next task was to gather the financial strength to realise a demonstration project of hydrogen and fuel cells in the transport sector. Pall Kr. Palsson of the New Business Venture Fund became the first chairman of the Board, with board members from the three international partners: Bjorn Sund from Norsk Hydro, Phillip Mok from Daimler Benz and Harald Schnieder from Shell Hydrogen. The Icelandic holding company Vistorka (EcoEnergy) held 51 per cent of the initial shares and I myself was a board member for Vistorka together with Palsson.

Icelandic New Energy employed a young and active person, Mr. Jon Bjorn Skulason as general manager which turned out to be a wise decision at that point. Jon Bjorn had a background as an economical geographer who had received his training at Simon Fraser University in Vancouver and had been working on concepts of battery driven cars and as a new business consultant for the municpaility of Keflavik near the capital. Jon Bjorn, assisted by Maria Maack, later employed as environmental manager of Icelandic New Energy, together with Hjalti Pall Ingolfsson, a mechanical engineer, have really proven a great asset for the nascent corporation and deserve a lot of credit for their hard work. Gradually there was a small dedicated group in Iceland, Bragi Arnason, usually called the father of the Icelandic Hydrogen movement, Jon Bjorn, Maria Maack, me and the board members of Vistorka as well as Hjalmar Arnason. This group of just over a handful of people was the executive group of the Icelandic hydrogen energy economy for over half a decade. This would all be in vain if it were not for the unremitting encouragement from the Ministry of Energy, Trade and Industry led by Minister Valgerdur Sverrisdottir and other ministers to follow in her footsteps.

The European Commission granted the project which was given the name Ecological City TranspOrt System (ECTOS) about three million Euros and the partners put another four million Euros into the project. A lot of work was put into the preparation for the first project. It became my task, in the new role as Chairman of the Board to inaugurate a commercial hydrogen refuelling station built in Reykjavik and put into operation three fuel cell buses imported to Iceland for demonstrations

It seemed ideal for the island to use hydrogen produced on-site from the available renewable electricity to power the car fleet. Furthermore, only about a half a dozen of stations on the 1400 km long ringroad around the island would create a customer acceptable minimum infrastructure.

The Icelandic project plans initially sparked off enormous reaction in Europe. Apparently, nine cities in Europe wanted to perform the same kind of test as we were planning in Iceland and had consulted DaimlerChrysler to operate their buses as a part of varying hydrogen energy chains. This movement finally materialised with the crea-

Figure 74 The ECTOS bus at the hydrogen station in Iceland.

tion of a much larger project involving 27 buses, called Clean Urban Transport for Europe (CUTE), taking place in nine European cities.

The overall objective of ECTOS project launched in 2001 has been to implement a demonstration of the state-of-the-art hydrogen technology by running part of the public transport system with fuel cell buses within Reykjavík. The energy chain will be next to CO_2 free, because domestic geothermal and hydro-powered energy sources will be used to produce hydrogen by electrolysis. A fuel station comprising production, compression, storage and filling was built in 2003 and produced very pure hydrogen for the ECTOS fuel cell buses. The fuel cell buses are refueled daily at the hydrogen station, therefore all transportation and extra handling is minimized and safety measures optimal. The main research objectives concern the socio-economic and environmental aspects and forecasting for an entire shift from fossil fuels to a hydrogen-based modern society.

Within ECTOS, environmental, social and economic aspects of using hydrogen as a fuel have been studied. The technological performance forms the basis for a comparative assessment. The ECTOS project is the first step in a foreseen transition to a hydrogen-based economy in Iceland.

The goals of ECTOS were learning by doing and gaining real experience from using hydrogen as a fuel.
- Experience of establishing a new fuel station and interactions with the regenerative electric supply system.
- Find the estimated costs and timeframe of integrating a new distribution system for hydrogen infrastructure.
- Contribution to clean air by creating a CO_2 neutral public transportation system.
- Map the public acceptance of using a renewable energy source to power the transport system, locally made and independent of fossil fuel supplies.
- To evaluate the life-cycle impacts of the hydrogen equipment (buses and the filling station) and the well to wheel impacts of the Icelandic hydrogen fuel chain.

Figure 75. Bragi Arnason, father of the Icelandic hydrogen movement. Source: Icelandic New Energy.

Another important project was run with support from the EU. That is the EURO-HYPORT project with the aim to examine the feasibility of exporting hydrogen from Iceland to the European continent. The investigation was in a number of steps: Hydrogen production methods in Iceland (cost, emissions and other factors), form of hydrogen (liquid, gaseous, etc.) and efficiency of hydrogen production. The cost of transport of hydrogen was studied in three different size scenarios. Basic infrastructure needed in Reykjavik was assessed. Finally, the Icelandic New Energy team produced an educational compact disk in association with the European Commission.

One of the blessings of the ECTOS project in Iceland is that it always enjoyed the support of the Icelandic people, who in polls showed up to 90 per cent support for the hydrogen energy economy. The act of "learning by doing" has been the hallmark of the

Figure 76. The first buses arrive in Iceland in 2003. From left to right: Hjalmar Arnason (temporarily with a broken leg), Jon Björn Skulason, Mayor Thorolfur Arnason and Thorsteinn I. Sigfusson. Source: Icelandic New Energy.

project and a lot of things have been learned. It is expected that the next generation of hydrogen buses will benefit greatly from the ECTOS and the CUTE experience.

University of Iceland, Technological Institute and ministries of Energy and State are all working hard in the area of hydrogen research. Icelandic New Energy is generally seen as its spin-off. In recognition of Bragi Arnason´s pioneering work, the combined effort within the University is named "Bragastofa". Some of the effort concerns fundamental studies of the chemistry of the various hydrogen bonds in compounds led by Prof. Hannes Jonsson, studies of nanostructural aspects of hydrogen storage by Sveinn Olafsson, studies of socioeconomic nature led by Örn D. Jonsson and Maria Maack and studies of geothermal hydrogen energy production and storage systems led by Thorsteinn I Sigfusson and Bragi Arnason. The projects involve a dozen graduate students and a number of staff in addition to the ones mentioned above.

Work on sodium borohydrides and cooperation with the university in the area of geothermal hydrogen as well as a roadmap building for the Ministry of Energy are also in progress. The Ministry has been very active in the area of hydrogen. The Ministry of Foreign Affairs is participating in the steering work of the International Partnership for the Hydrogen Economy.

Throughout the whole life-time, the Icelandic projects have enjoyed the support of three successive prime ministers: David Oddsson, Halldór Ásgrímsson and Geir H. Haarde. The President of Iceland, Olafur Ragnar Grimsson, has been a steadfast supporter of the work since its early phases.

The government of Iceland has in 2006 decided to actively participate in the second phase of a demonstration fleet project in Iceland. This time, it involves over 30 fuel cell vehicles as well as hydrogen powered internal combustion engine cars that will run until 2010. Furthermore, Iceland will test hydrogen at sea by powering an auxiliary generator of a tourist boat with a fuel cell system in stead of diesel combustion engine. This second step towards a hydrogen society in Iceland is called "Sustainable Marine and Road Transport, H2 in Iceland", SMART-H2. The first car in the project was inaugurated by Energy Minister Össur Skarphedinsson and Mayor Vilhjálmur Vilhjálmsson in July 2007.

Every year hundreds of journalists and documentary producers visit the Icelandic hydrogen team. Throughout the winter, the University team and Icelandic New Energy run an information/education seminar and already hundreds of individuals have completed the course. International summer schools have been held in Iceland and the University staff have been giving lectures and seminars all around the globe. The phrase "hydrogen tourism" has been rightly coined!

If the plans for developing a hydrogen energy economy in Iceland become a reality, the country can approach almost 100 per cent renewability. At the time of this writing, we are 8 years into the plan for the estimated 50 years conversion of Iceland. The phase is still "demonstrations". An island economy based on renewables like Iceland is in many ways ideal for such a transformation. The fate of this conversion will depend upon many factors. One of the most important factors is international cooperation. A totally renewable energy island could emerge before the middle of the 21st century.

India, the second wildcard

India addresses hydrogen both in terms of energy security as well as environmental protection. The government sees its hydrogen energy economy evolve from the intermediate bio-fuel and synthetic fuel based transportation as well as electric and hybrid vehicles into the "ultimate objective" of "environmentally friendly and carbon free, hydrogen vehicles and power generation".

India imports more than 72 per cent of its oil demands and the oil imports are expected to significantly rise in next fifteen years. However, renewable energy is abundant in India, indicating a major potential to contribute to energy security.

Dr. S.K. Chopra, a principal adviser to the Indian government here has pointed out that large areas of the country have no access to electricity, but can become a part of a de-centralised power system based on hydrogen energy. In this context, he expects that hydrogen and fuel cell vehicles could progressively take the place of petroleum based vehicles, specifically two and three wheelers that are now so popular in India.

The Ministry of Non-Conventional Energy Sources has been actively involved in hydrogen research and development over two decades. One of the highlights is a broad programme for research, development and demonstration on different aspects of hydrogen energy including production, storage, transportation application and power generation.

A considerable progress was made when India established the National Hydrogen Energy Board (NHEB) in October 2003 to bring about the accelerated commercialisation of hydrogen energy technologies. A Steering Group on Hydrogen Energy formed under the NHEB and drafted the *National Hydrogen Energy Road Map*. The road map has highlighted hydrogen production as a key thrust area of research, development and demonstration. For example, hydrogen production from nuclear energy, coal gasification, biomass, biological and renewable energy-based methods have been identified for further development to provide low cost and preferably carbon free hydrogen. In the area of hydrogen storage, improvement of storage efficiency and cost reduction were emphasized. The road map has been accepted by the NHEB and is being processed for its implementation.

The road map has also identified two major initiatives with goals and targets up to 2020 – Green Initiative for Future Transport (GIFT) and the Green Initiative for Power Generation (GIP). The former aims to develop and demonstrate hydrogen powered internal combustion engine (ICE) and fuel cell based vehicles ranging from small two/three wheelers to heavy vehicles through different phases of development. The road map envisages that by 2020 approximately one million hydrogen fuelled vehicles, of which about 75 per cent are expected to be two and three wheelers, would be running on the Indian roads. In the initial years, the automotive sector will focus on hydrogen powered ICE. The industry is expected to make a switch over to fuel cells as the cost is reduced and improvements in fuel cells are achieved. Meanwhile, GIP aims to develop and demonstrate decentralised power generating systems based on hydrogen powered internal combustion engine/turbine and fuel cells, ranging from small watt capacity to mega watt size systems through different phases of technology development and demonstration.

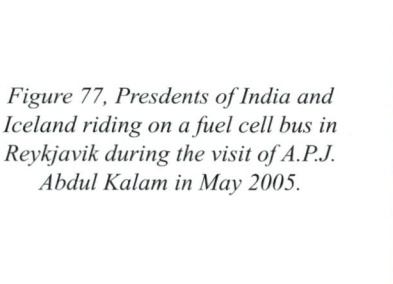

Figure 77, Presdents of India and Iceland riding on a fuel cell bus in Reykjavik during the visit of A.P.J. Abdul Kalam in May 2005.

The first Indian hydrogen fuelling station was opened in Faridabad, north of New Delhi in October 2005. This facility has been used for dispensing a hydrogen-compressed natural gas blend as well as pure hydrogen. Similar facilities for hydrogen production and dispensation are expected to be set up in New Delhi and other major cities in the future.

Additionally, significant efforts are being made to start a pilot plant for coal gasification as well as a demonstration plant set up for a biological route for hydrogen production. The research into metal hydrides for hydrogen storage has resulted in hydrides with 2.42 wt per cent storage capacity for ambient conditions developed in 2005. India is also working on carbon nanostructures for hydrogen storage as well as the development of hydrogen ICE and PEM fuel cell for stationary applications and automobiles.

Italy and the renaissance of hydrogen

Hydrogen research has a long standing tradition in Italian universities. At the beginning of the 1990s, some projects for the hydrogen production from renewable sources and its use in internal combustion engines were implemented. The fuel cell activities started in the1980s and focused on the development and demonstration of PEM fuel cells for stationary and transportation applications and molten carbonate fuel cells (MCFC) for on-site and distributed generation.

Recently, Italy has seen a revolutionary public movement towards hydrogenisation. Italy held the presidency of the European Commission during a time when hydrogen was a key point of emphasis in the energy politics in Brussels. The meeting of Environment and Energy Ministers of 30 European countries in Montecatini, Italy in July 2003 provided an important forum to consolidate energy and environmental considerations in the continent´s policies. From this meeting onwards, de-carbonisation became an ever increasing issue in political work in Europe.

Ministry of Research and University and Ministry of the Environment supported a

National R&D Programme on Hydrogen and Fuel Cells in 2003, which was outlined in the framework of the *National Research Plan* (PNR). A number of initiatives was started at research institutes and industries in the framework of the National R&D Programme in 2005. The programme has funded 14 projects with about 90 million Euros through the Special Integrative Fund for Research (FISR) for the period 2005-2008. To be specific, the projects on hydrogen supported by the FISR include the development of technologies for hydrogen production from renewable sources and from fossil fuels. This is done with CO_2 capture and sequestration and the development of systems for hydrogen storage and for hydrogen use in the transportation and distributed generation. The FISR projects on fuel cells aim to improve performances and reduce cost through the development of materials. Through the FISR programme more than 100 research groups gathered including ENEA (National Agency for Energy, Environment and New Technologies), CNR (National Research Council) and universities in cities such as Milan, Turin, Messina, Salerno and Palermo.

Italian regions have also shown considerable interest in hydrogen as a tool for creating sustainable energy and transport policies and have started financing initiatives to support demonstrative projects. Especially, scientific and technological parks across the country have been used as research labs on hydrogen technologies. Here are examples of the hydrogen parks. In Turin, Piedmont, Hydrogen System Laboratory (HySyLab) actively develops the entire supply chain from hydrogen production through storage and applications into environmental modeling. In Milan, Lombardy, hydrogen production from natural gas and its utilisation in stationary and transportation application are tested through the Bicocca Project, an urban hydrogen laboratory that currently features three conventional cars that are converted to hydrogen. In Mantova, Lombardy, RD&D infrastructures for alternative motor fuels are developed through the Zero Regio Project. In Marghera, Veneto, integrated project for stationary and automotive applications will be carried out, such as coal gasification, MCFC fuel cell integrated with cogeneration plant, hydrogen pipeline in urban area, hydrogen refuelling stations, through the Hydrogen Park Project. In Abruzzo, renewable energy system is installed in combination with hydrogen production and fuel cell application. Besides these, Florence and Arezzo in Tuscany also operate hydrogen parks.

Figure 78. Fiat Panda Hydrogen Fuel Cell Vehicle from Italy.

Several Italian companies are also heavily involved in hydrogen utilisation related to transportation and fuel cells. Fiat Powertrain develops fuel cell vehicles powered with hydrogen and fuel cells for a hydrogen urban vehicle. In addition, Ansaldo Fuel Cells develops MCFC stacks and a couple of Megawatt demo unit. Nuvera Fuel Cells develops 5-25 kW electric power modules for industrial and commercial use and 4kW CHP (combined heat and power) systems using PEM fuel cells. Arcotronics Fuel Cells develops CHP systems for domestic applications and PEM fuel cell stacks with power in the range 500W to 50kW.

In 2003 the Italian Hydrogen and Fuel Cell Association, H2IT, was established including the main power and gas companies as ENEL and Edison, ENI/AGIP, SAPIO, SOL, Air Liquide, Linde and Italian Regions among over a 100 members becoming the biggest hydrogen association of Europe. H2IT is participating as partner for dissemination in 8 EU projects and developed a Roadshow "Hydrogen hits the roads!" to visit local authorities, schools and Chambers of Commerce with information material and small hydrogen applications. H2IT is coordinating the HYFED6 secretariat that is coordinating the collection of educational programmes, organisation and university courses on hydrogen and fuel cells.

Recently five Regions interested in integrating the use of hydrogen into their energy and transport policy, prepared a proposal for the Italian Ministry of Environment pointing to the role of hydrogen/methane mixtures in facilitating local infrastructure development. The Region of Lombardy, that published the first regional "Hydrogen programme for the hydrogen vector" in Italy in November 2006, includes substantial financing for refuelling stations and cars that run on mixtures. Fiat announced at the beginning of 2007 that their new Fiat Panda that runs on methane will also be adapted to mixtures of 30 per cent by volume hydrogen.

Japan: protonics added to electronics

What Japan achieved over the past decades with the utilisation of the electron with the development of consumer electronics and microelectronics – could be reincarnated in their harnessing of the proton with hydrogen technology. In my frequent visits to Japan I have been able to get a closer understanding of the Japanese industrial genius and I am convinced that Japanese technology will prove a strong challenger to the world hydrogen development.

Japan depends heavily on energy imports for over 80 percent of its energy supply, implying that the importance of hydrogen and fuel cell technology is increasing, specially due to the recent rise in oil prices and political instability in the Middle East.

Prime Minister Junichiro Koizumi gave a powerful speech to the National Diet of Japan in 2002 setting the future tone for hydrogen: "The fuel cell is the key to opening the doors to a hydrogen energy economy. We will aim to achieve its practical use as a power source for vehicles and households within three years". Interestingly, in April 2005, the Prime Minister's residential quarters installed the world's first commercial household fuel cell unit produced by Panasonic and Ebara-Ballard.

The Ministry of Economy, Trade and Industry (METI) has invested in fuel cell

Figure 79 The Japanese Hydrogen Highway. Source: JHFC.

development since 1981 and METI's scenario for the practical realisation and dissemination of fuel cells was unveiled in 2001. The government strategy is based on a three-stage commercialisation plan : Introduction Stage (2005-2010), Diffusion Stage (2010-2020) and Penetration Stage (2020-2030). In the first stage, the introduction of vehicles will be accelerated with the gradual establishment of fuel supply system. In this context, Japan Hydrogen and Fuel Cell Demonstration (JHFC) will examine the effectiveness, environmental impact and the safety of fuel cell vehicles and refuelling stations. The strategy also set numerial targets for hydrogen and fuel cells technologies, for example, 5 million fuel cell vehicles and 10 GW for the total power generation by stationary fuel cells by 2020. To meet these targets, METI has allocated an annual budget of over 30 billion JPY since 2003. The scale of this support is by all measuers huge.

In 2005, the New Energy and Industrial Technology Development Organisation (NEDO) as METI's funding organisation started issuing the roadmap for hydrogen and fuel cells technologies on an annual basis, including Proton Exchange Membrane Fuel Cells (PEMFC), Direct Methanol Fuel Cells (DMFC), Solid Oxide Fuel Cells (SOFC) and related hydrogen technologies. The roadmap set up a number of phases until full dissemination by 2020 to 2030 and shows numerial targets for basic performance, durability and costs for each phase.

Japan already tested 58 fuel cell vehicles before September 2005 through JHFC programme which comprises fuel cell vehicles (FCV) demonstration study and a similar emphasis on hydrogen stations. Global motor companies such as Toyota, Nissan, Honda, DaimlerChrysler, GM, Hino, Mitsubishi and Suzuki joined the JHFC programme while most of fuel cell vehicles were domestically produced. Moreover, ten hydrogen stations were built to support road tests in Tokyo and Kanagawa area and two additional hydrogen stations were installed to support fuel cell buses of 2005 Expo in Aichi. In April of 2004 a small hydrogen fuelling station was inaugurated on the island of Yakushima about 60 km from the southernmost tip of Kyushu in Japan. The station uses a PEM electrolyser and obtains electricity from a hydroelectric power facility on the island.

A large scale stationary fuel cell demonstration project has been in progress, with 400 individual polymer electrolyte fuel cells (PEFC) tested and about 1300 fuel cells installed in March 2007. The outlook of residential stationary fuel cells is favourable in Japan, where the price of power is very high for residential power. The governmental funding for hydrogen and fuel cells in Japan is close to 0.3 billion USD annually.

In recent years, Japan has established two national laboratories for research and development of hydrogen and fuel cell technology. The ambitious Fuel Cell Cutting Edge Research Centre, also known as FC-Cubic, was opened in April 2005 focusing on basic scientific knowldge to drive innovation in close collaboration with industry, public and academia. Likewise, the Research Centre for Hydrogen Industrial Use and Storage, also known as HYDROGENIUS, opened its doors in the Kyushu University in July 2006, dedicated to performing fundamental studies on fatigue and fracture of materials at ultra high pressure. It is the hope of Japan that it becomes a unique international centre of excellence for hydrogen technologies.

Korea joining the forefront

Hydrogen has been on the list of energy policy agenda in The Republic of Korea since 1989. The efforts of promoting hydrogen and fuel cell technology in Korea is closely linked with domestic energy supply and demand situation : Despite few indigenous energy resources, Korea is the tenth largest energy consumer in the world, importing about 97 per cent of its energy from foreign countries. The energy consumption fuels among other activities a fast growing automobile market as well as linked industries.

To ensure a stable supply of energy and to establish an environmentally-friendly, low-carbon energy system, Korea plans to increase the share of new and renewable sources of energy in total primary energy supply from 2.13 per cent in 2005 to five per cent by 2011. To achieve this challenging goal, Korea selected hydrogen fuel cell, photovoltaics(PV) and wind power as the three major areas with market potential. In addition, the hydrogen fuel cell was also recognized as one of the top ten economy growth engines for the next decade.

In December 2003, the government announced the Basic Plan on Developing and Disseminating New and Renewable Energy Technologies to promote the development and dissemination of new and renewable energy technology. This plan has become the foundation for establishing Korea's ambitious Hydrogen and Fuel Cell Programme.

The R&D funding for hydrogen and fuel cells is provided by the Ministry of Commerce, Industry and Energy (MOCIE) and the Ministry of Science and Technology (MOST). MOCIE is mainly engaged in short- and mid-term projects and the development of industrial application technology. For transportation applications, MOCIE launched the 80kW Proton Exchange Membrane Fuel Cell (PEMFC) Vehicle Programme and the 200kW PEMFC Bus Programme in 2004 and 2005 respectively. Hyundai-Kia Motors are the main contractors for the two projects, providing half of the project funding. At the same time, three hydrogen refuelling stations will be opened in Seoul, Incheon and Daejeon during 2006 and 2007 to support the deployment of hydrogen fuel cell vehicles across the nation. These stations are expected to provide hydrogen fuels made from liquefied natural gas, liquefied petroleum gas and naphtha.

Korea also works on residential power generation, with both PEMFCs and Solid Oxide Fuel Cells (SOFC) being considered for the commercialisation of power genera-

Figure 80. Korean hydrogen vehicle foundations shown by KIA Motors. Source: KIA.

tion. Two Korean companies - GS Fuel Cell and Fuel Cell Power - have developed their respective PEMFC systems for residential power generation. For industrial power generation, Molten Carbonate Fuel Cell (MCFC)-based distributed power generators are being primarily considered for industrial power generation. Korea Electric Power Corporation (KEPCO) plans to develop the 250 kW MCFC system by 2009. In addition, the development of both types of the PEMFC and Direct Methanol Fuel Cell (DMFC) is strongly considered for portable applications. Currently, Samsung SDI is developing the 50W PEMFC system for laptop computers and LG Chem is developing the 50W DMFC system for other portable devices. Education is also an important element of MOCIE programme. In order to meet the demand for R&D manpower for hydrogen and fuel cell technologies, Korea started various hydrogen education programmes in 2005, including short-term re-education to industries by Core-technology Research Centres, the establishment of specialized graduate schools and the selection of best labs for in-depth research.

Meahwhile, the Ministry of Science and Industry, MOST, has funded long-term projects and development of basic technology. The Ministry announced to support an impressive Nuclear Hydrogen Programme in the period 2004-2019 and approximately 1 billion USD will be put into the programme. The target is to complete the development and demonstration of the nuclear based hydrogen production technology and cover about 20 per cent of the energy demanded in the transportation sector in the 2020s. Besides, the Ministry has supported hydrogen production, storage and utilisation technology through its 21st Frontier Hydrogen Energy R&D Programme.

It will be exciting to follow the development of hydrogen in the fast moving technological society of Korea in the near future.

Norway and the Nordic pioneers

Hydrogen has a long standing tradition in Scandinavia. Originally, some of the pioneers in water electrolysis were associated with the Norwegian company Norsk Hydro. The hydroelectric harnessing in Norway was closely associated with production of fertilizers which were based on nitrogen from ammonium nitrate derived from ambient nitrogen and hydrogen from electrolysis of water. Norwegian scientists and engineers have influenced hydrogen research and development in the Nordic countries to a considerable extent.

The veteran Norsk Hydro performed industrial electrolytic hydrogen production during the period 1920 – 1990. After the switch to hydrocarbon-based hydrogen production, the competence in water electrolysis was maintained and exploited by the daughter company Norsk Hydro Electrolysers. Christopher "Toffen" Kloed, who was in charge of this company in the 1990s, has been a keen promoter of hydrogen as an energy carrier. He was one of the pioneers of the establishment of Icelandic New Energy together with Bjorn Sund of Norsk Hydro. The Norwegian technology was well known in fertilizer production in Iceland.

Another pioneer in Norway, Knut Andreassen, has been a prominent expert on water electrolysis. Around 1990, he led a comprehensive study called NHEG (Norwegian Hydro Energy in Germany), concerning possible hydrogen production with hydropower in Norway and export as liquid hydrogen to markets in Germany. The study was published at the 9th World Hydrogen Energy Conference in Paris 1992. Research in water electrolysis is a long time specialty of the technical university NTNU in Trondheim, and Prof. Reidar Tunold has been a leading figure in this area. Thorstein Vaaland of Oslo University and Agder College in Norway have also been leading researchers in this area.

The physics department of the Norwegian Institute for Energy Technology (IFE) has been engaged in metal hydride research since the 1950s. In the first decades this was largely an academic exercise, but in the 1970es interest arouse in the use of metal hydrides for hydrogen storage. In 1977 an international symposium entitled "Hydrides for Energy Storage" was held in the mountain town of Geilo, Norway. The proceedings, edited by Arne F. Andresen and Arnulf J. Maeland, were published by the International Association for Hydrogen Energy. Maeland is living in the USA, but has always been an active supporter of the metal hydride research in Norway. Arne Andresen, who died in an accident in the late 1980s, was also a true pioneer in this area. He was in charge of a large research project in the period 1978 – 1984, concerning magnesium hydride for hydrogen storage. This project was financed by Norsk Hydro and also involved the Risø National Laboratory in Denmark. There has also been cooperation with Sweden. Prof. Dag Noreus of Stockholm University is one of the metal hydride pioneers and has contributed especially to the development of metal hydride batteries. In the 1990s the work at IFE was at a rather low level due to budget limitations, but since 2000 there have been great expansions, and the researchers at IFE, teams led by Björn Hauback and Volodymyr Yartys and many others, are now regarded as one of the leading groups in this area on the international forum.

Based on general competence in renewable energy and fuel cells, IFE got engaged in studies of SAPS (Stand Alone Power Systems) in the late 1980s. Bjørn Gaudernack was involved in several of these like a study of a wind power, water electrolysis and fuel cell plant. He was also engaged in the SAPHYS Project, a JOULE-II project concerning a pilot plant for solar hydrogen production in Casaccia, Italy. He also participated in the planning of a similar small pilot plant for Agder College at Grimstad, which became operative in the late 1990s, and he was also a strong advocate of hydrogen production from natural gas with CO_2 sequestration.

The Nordic countries have a tradition for cooperation in research. Within the Nordic Energy Research Programme, an activity on fuel cells and, later, electrochemical energy conversion was included from 1991. Bjørn Gaudernack was in charge of this activity during the 1990s. On his initiative, a Nordic Working Group on hydrogen as an energy carrier was established in 1992. A leading figure in the group was Bragi Árnason from University of Iceland. Participating in the group were also Dag Noreus from Stockholm, Lotte Schleisner from Risø, and Peter Lund from Helsinki Technical University. The group performed a study of the potential for "clean" hydrogen production in the Nordic countries, and concluded that the potential was formidable. From 1992 onwards, Gaudernack was engaged in the IEA Hydrogen Programme as Operating Agent for an Annex on photo-production of hydrogen. Prof. Peter Lindblad from Uppsala University was a strong contributor to this work. Trygve Riis from the Norwegian Research Council was a member of the Executive Committee of the Programme and later became chairman of this committee. The Norwegian Hydrogen Forum was established in 1994, on the initiative of Christopher Kloed (Norsk Hydro Electrolysers), Bjørn Gaudernack (IFE) and Tor Sætre (Agder College). Christopher "Toffen" Kloed was the first leader of the Forum.

Based on the tradition described above, the Norwegian government appointed in June 2003 a National Hydrogen Commission to work out a broad rd&d programme covering production, storage, distribution and end use of hydrogen both in the transport sector and in stationary applications. The Commission that comprised the members from industry, governmental institutions and NGOs published its report *Hydrogen as the Energy Carrier of the Future* in June 2004 and recommended a national hydrogen rd&d programme. As a follow-up of the report, Minster of Transport and Communications, Torid Skogsholm and Minister of Petroleum and Energy, Thorild Widvey announced the *Norwegian Hydrogen Strategy* in August 2005. In the strategy, utilisation of natural gas resources, industrial development, environmental benefits and participation in the forefront of international research were presented as four factors that support investment in hydrogen. In addition, the national hydrogen platform was established to encompass current measures and the funding related to hydrogen and to unite hydrogen activities under a single umbrella. This platform went into operation in January 2006 with subsequent establishment of Strategic Council and Action Plan in December 2006 and enjoys the support of a broad range of government agencies, corporations and institutions in Norway.

While Norwegian companies and research groups are participating in several demonstration projects in Europe such as CUTE and ECTOS, Norway is also operating significant demonstration projects of its own. The major project at present is the forementioned Utsira (Utility Systems in Remote Area – the name is a perfect example of a project name adapted to a place name!) project. Norsk Hydro has established a small renewable hydrogen society at the island of Utsira in the North Sea with co-funding from the government. The Utsira project is an autonomous system with wind turbines, electrolyser, storage system, hydrogen combustion engine and fuel cell as the basic components. The project was opened in July 2004 and the demonstration is expected to be extended a few years

Norway also enjoys a leading position in the hydrogen production from electroly-

Figure 81. The TH!NK hydrogen and fuel cell car at the onset of a Norwegian – Danish cooperation.

sis. In fact, Norway has large renewable energy sources and hydro-power accounts for over 99 per cent of the country's domestic production of electricity. Norwegian industrial companies have more than 75 years of experience in the hydrogen production by electrolysis and the Norsk Hydro produces electrolysers for the world market. Thus, further development and cost reduction of electrolyses, including PEM electrolysis is expected to be a priority area in Norway.

Another groundbreaking project is the HyNor, which is to demonstrate the commercial viability of hydrogen production and use in the transport sector. This project involves a real life implementation of hydrogen infrastructure along a route of 580 km from Oslo to Stavanger in the period 2005-2008. This will include various hydrogen production technologies and 5 filling stations. Hydrogen will be obtained from electrolysis, biomass, a petrochemical plant and natural gas steam reforming. The end users will be buses, taxis as well as private cars. The first hydrogen filling station opened on the outskirts of Stavanger in August 2006. Due to Norway's long coastline and substantial competence and activities in the marine and shipping industry, Norway also explores the possibilities to develop and demonstrate hydrogen fueled vessels, for example in ferries.

Finally, the TH!NK hydrogen, a partnership between Raufoss Fuel Systems and Think Nordic AS, develop a prototype fuel cell/electric hybrid vehicle manufactured in Norway. The car, called TH!NK hydrogen, will provide the freedom to charge the batteries from a plug in the wall or fill the tank with hydrogen. With a hydrogen fuel cell range extender, the driving range can reach as far as 300 km. The project is split into three phases from 2005 to 2007 and a small series of hydrogen electric cars will be delivered in 2007.

Turning to the R&D efforts of Norway, the Norwegian research institutes have substantial activities in the hydrogen field such as hydrogen storage. Accordingly, very high hydrogen density has been achieved in a new type of solid metal hydride. Norwegian research and demonstration will be more focused on with the establishment of HYTREC (Hydrogen Technology Research Centre). This centre will be based in Trondheim and embrace research facilities as well as practical demonstrations of using hydrogen for transport, electricity generation and heating. The plan for establishing

HYTREC was drawn up by power utility Statkraft, the oil and gas company Statoil and classification society Det Norske Veritas (DNV). The centre has been operational since early 2007.

Other Scandinavian countries are active in hydrogen research as can be seen in various sections of this book. We have limited our narrative to the direct IPHE countries but will also mention a few other Scandinavian highlights briefly.

In Denmark, the Risoe National Laboratory near Roskilde is a remarkable and innovative centre of hydrogen expertise. As an example I can mention that the number of fuel cell patents from Risoe is one of the highest in the world. For many years Risoe has cooperated with Haldor Topsoe company, one of the world leaders of catalyst technology. This cooperation has recently led to the formation of a new Danish company focusing on SOFC fuel cells. The Danes achieved a great success in wind energy conversion where they combined their scientific skills with clever marketing. In hydrogen my anticipation level for Danish success is great.

Nearby at the Roskilde University Centre there is an active group on hydrogen associated with Bent Soerensen a theoretical physicist who has devoted his creative time to hydrogen research and textbook writing.

At the Danish Technical University near Copenhagen, hydrogen plays a role in a number of interesting projects. The most notable one is perhaps the "hydrogen tablet" development. The tablet is a metal ammine domplex that stores 9.1 per cent hydrogen by weight. The storage is completely reversible and by adding an ammonia decomposition catalyst hydrogen can be delivered at temperatures below 347°C. Claus Hviid Christensen Professor of Chemistry has been in charge of this work.

In Herning in the Jutland heath close to the North Sea coast there has been an interesting development in the area of hydrogen with the work of the Hydrogen Innovation & Research Centre. The hytopia of their vision has been coined into a trademark H2PIA where a number of companies and institutions join forces in founding a new thinking in hydrogen. One of the most notable members of the Herning team is a small and innovative company H2Logic founded by young experts in Herning led by Jacob Hansen. The company is active in the preparation of the future fuel cell powered transportation systems. The Danish Hydrogen Association under the chairmanship of Jan Hovald Petersen has been operating since 2005.

We continue our trip through Scandinavia and come to Finland. Here, the VTT Technical Research Centre of Finland is doing important work on hydrogen, mostly focused on hydrogen from biomass and SOFC fuel cells as well as industrial applications of PEM fuel cells. As in many Scandinavian countries, much of the hydrogen work in Finland is linked to the ministerial Nordic Energy Research in Oslo. Many of the hydrogen projects are related to universities like the Helsinki University of Technology, The Technical University of Tampere and Aabo Academi.

It will also be worth while to observe the progress of research and development in the Finnish company Vartsilla. The company is a world leader in engines for ships and vessels and have been including hydrogen and fuel cells in their development portfolio.

Towards the end of our brief Scandinavian tour outside Norway, we come to Sweden. We already got some insight into Swedish hydrogen research in our mentioning of artificial photosynthesis and photobiological hydrogen production as well as the Chrisgas project earlier in this book. The Icelandic hydrogen demonstration project enjoyed Swedish expertise in the traffic studies with experts from the Vinnova organisation that were involved.

One of the prominent names in Swedish hydrogen research is Erik B. Karlsson of Uppsala University who was a pioneer in for example metal hydride research. I must mention that among his students was one of the Icelandic pioneers in hydrogen research, Björgvin Hjörvarsson who is based at the Angstrom Laboratory in Uppsala. Most of the Swedish and Scandinavian hydrogen research is of very high quality and would deserve a much thorough description which is outside of the present scope.

Russia and the increased strong commitment to hydrogen

In the former Soviet Union, hydrogen enjoyed considerable interest in science and technology early in the 20th century. At the beginning of the century K.E. Tsiolkovsky found the basis of the jet motion and proved the possibility of inter planet flights with the help of rocket engines using hydrogen fuel which was to become a hallmark of human development.

In the 1920s and 30s a number of demonstration projects involving hydrogen were performed. For example wind energy based hydrogen production was already tested before mid century.

Stanislav Malyshenko, one of the hydrogen pioneers of modern Russia has told me some of the interesting history of Russian hydrogen where the Second World War was at a focal point. According to Malyshenko, during the Leningrad siege in 1941, Lieutenant-engineer B.I. Shelish converted a large amount of automobile engines GAS-AA, which were used for driving the aerostat winches to hydrogen engines using hydrogen-air mixture from the aerostats that had lost their floatage. Aerostats are lighter than air and are based on floating in air. The mixture contained 15-20% of air and the backfire of their turbines could lead to aerostat explosion. To prevent the explosion, Shelish used water lock, installed before the engine and a number of other measures using other improvised means. Beginning from 1942, hydrogen from the aerostats, which had lost their floatage, was used also by Moscow antiaircraft service. During the war, more than 400 car engines for barrage balloon winches used hydrogen as a fuel.

Just like the rest of the world, the oil crisis in the 1970s sparked off much interest. A Volga car running on a mixture of hydrogen and gasoline was tested in the beginning of the 1980s. The world's first hydrogen powered mini bus, Kvant RAF, was tested in 1982 in Russia.

One of the most impressive pioneering work in Russia was the development of the TU 155 aircraft with hydrogen-gasturbine powered engines. A unique cryogenic hydrogen complex was developed at the Baikonur cosmodrome. This was followed by alkaline fuel cell development for space exploration use. Simultaneously the development of metal hydride storage system was an important area of research.

The Ministry of Industry, Science and Technologies of the Russian Federation performed project in hydrogen technologies in the timeframe 2002-2006. PEM fuel

Figure 82. The Tupulev 155 Russian airplane, which was flown on hydrogen. Source V. Tatarintsev.

cells are studied at the Physics and Energy Institute and the Kurchatov Institute, the old nuclear knowledge base in Moscow. The Russian Academy of Sciences has worked on high temperature fuel cells. The space company Energiya is working on a new generation of alkaline fuel cells. At the Moscow motor show in 2003 the company showed a VAZ automobile powered by an alkaline fuel cell.

Russian hydrogen research today is focused on a range of subjects. Russia is becoming one of the fossil gas bases in the world and provides countries in Europe with natural gas. Highly efficient new processes of H_2 and syn-gas production from hydrocarbon material (natural gas, methane, propane-butane mixture, alcohols, liquid fossil fuel) on the base of plasma catalysis is a novel technology which Russia is promoting.

Plasmachemical technologies are being developed for large-scale hydrogen production from methane and hydrogen sulfide (acid gas) in non-equilibrium SHF – discharges.

Technology for natural gas thermal decomposition and waste wood processing yielding hydrogen and carbon (pyrocarbon of over 1.8 g/cm^3 density or other carbon material) is an important development item in Russia.

Among research and development items in Russia are new catalytic systems for hydrocarbon fuel processing and hydrogen production (including on-board and small scale devices) and advanced technologies for hydrogen purification and separation. Hydrogen production enjoys special attention and advanced electrolysis is among studied processes. High temperature nuclear technologies for hydrogen production from fossil fuels and water are an area of much interest in Russian hydrogen technology. Russia is interested and capable of large-scale hydrogen production with the use of nuclear energy, including the technologies based on high temperature nuclear reactors. The country has experience in the area of liquid hydrogen technology and safety which is closely related to the space programme.

Research and development in these fields are carried out by many institutes of Russian Academy of Sciences, by Russian scientific centres ("Kurchatov Institute", "Applied Chemistry" and others), by institutes and centres of Nuclear Energy Ministry (Minatom) of Russian Federation, Russian Aviation & Space Agency and universities.

In addition to about five universities offering graduate hydrogen education, two inter-university centres for joint research and training in hydrogen and fuel cell technology have been established in close cooperation with the Moscow power engineering institute and the Moscow based Institute of Chemical Technologies.

The Institute of People's Economy Prognostication of RAS has carried out socio-economic studies linked to the "Kurchatov Institute", Institute of High Temperatures and by others. Russia has expressed interest in comparative analysis of the social cost of the hydrogen fuel with respect to ecological impact on the whole cycle from production to end user.

Armed with the experience from a long space programme, Russia has wide interest in fuel cells and is for example working on Alkaline Hydrogen-Air Fuel Cells with removing of CO_2 from the incoming air. Then Russia is working on PEMFC with membrane, produced by the company JSC "Plastpolymer". Sophisticated work on Planar SOFC solid oxide fuel cells with yttrium stabilized zirconia electrolyte and various type catalysts is being done in Russia.

In this context, high-temperature membranes on the base of ZrO_2 and Y_2O with thickness 250-500 mm and 5-20 mm for different types of cells – electrolyte supported SOFC and anode supported SOFC. Also other types of high temperature membranes are investigated.

A strong movement in the field of non-precious metal catalysis for cathode and technologies for replacement of Pt with Pd is taking place among hydrogen scientists in Russia and is regarded as an important move to make the technology cheaper.

Hydrogen storage has a long standing tradition in Russia. Some of the areas of interest are: Low-cost and high efficiency metal hydrides for hydrogen storage and purification. Russian investigations are made of heat and mass transfer in dispersed metal-hydride media in sorption/desorption processes, including the case of the presence of admixtures in the incoming gas.

Traditionally Russia is a large producer of liquid hydrogen and advanced technologies of hydrogen liquefaction and LH_2 storage and transportation are an important part of the research portfolio. A technology of the production of cheap glass microspheres with further filling with hydrogen at pressure up to 100 – 200 MPa is being investigated as well as methods of hydrogen storage in carbon nanostructures, Fullerenes and similar.

In May 2006 President Putin signed a decree giving hydrogen energy technologies a critically important status in the development of the national energy economy. Hydrogen is a part of two new national programmes aiming for the time frame until 2012. One of the new programmes which is entitled "National Technology Basis for 2007-2011" has the aim to accelerate commercialisation of innovative technologies and new products including hydrogen technology. Large corporations like Nornikel, Gazprom and RAO, United Energy Systems, will contribute at least the same amount as the government to the projects. The governmental allocation to hydrogen is expected to rise from USD 15 million in 2005 to 25 million in 2007 which is one of the highest increases seen world wide in 2007.

United Kingdom, where much of the science was developed

The United Kingdom looks back upon glorious ages of hydrogen- related fundamental scientific discoveries. The pioneering achievements of Henry Cavendish and William Grove are good examples. At the biennial Grove Fuel Cell Symposium, now one of the traditional world conferences on the subject, experts from all around the world get together in the UK and reflect on the "state of the union".

The Energy White Paper released by the government in 2003 set ambitious goals for the UK's energy future, and the threat of global warming and climate change led the government to embark upon a path to reduce greenhouse gas emissions by 60 per cent by 2020. One important milestone in the area of hydrogen and fuel cell was laid when A Strategic Framework for Hydrogen Energy in the UK was published by E4tech, Element Energy and Eoin Lees Energy in December 2004. The Department of Trade and Industry (DTI) had commissioned the report in early 2004 to identify current UK expertise in hydrogen and find out what support was available. The key message from the analysis was that the use of hydrogen as a transport fuel offers significant opportunities for cost-competitive CO_2 reduction by 2030. After the release of the report, the Energy Minister, Malcolm Wicks MP, responded positively to its recommendations and announced a funding package of around 15 million pounds over four years for demonstrations of hydrogen and fuel cell energy technologies in June 2005. The government also accepted the recommendation for the establishment of a Hydrogen Coordination Unit (HCU) to facilitate the development and deployment of low carbon, low cost and secure hydrogen energy chains for transport. On the same day in June, Fuel Cells UK established itself as a fully fledged trade association and also published the *UK Fuel Cell Development and Deployment Roadmap*. This roadmap provided a framework of actions to help the UK optimise its response to the opportunities and overcome the challenges it faces. This work was funded by the DTI and involved an extensive process of consultation. UK's first Centre of Excellence for low carbon and fuel cell technologies Cenex was established by DTI in the spring of 2005. Public-private partnerships are an important part of the UK implementation. Cenex is an industry-led such partnership with the aim of assisting UK industry to build competitive advantage from the global shift to a low carbon economy. By encouraging the early market adoption of low carbon and fuel cell technologies in automotive applications, Cenex will assist UK industry to develop a supply chain capable of competing in global markets.

Meanwhile, the UK Research Council initiated the SUPERGEN programme to help the UK meet its emissions targets through a radical improvement in sustainable power generation and supply. Thirteen areas including hydrogen energy and fuel cells were identified as making a potential major impact on the UK's energy future and the first consortia were launched in November 2003. Currently, the UK Sustainable Hydrogen Energy Consortium (UK-SHEC) in SUPERGEN is taking a multidisciplinary approach to the many problems associated with turning hydrogen into a commercial fuel.

Figure 83. Peter Edwards of Oxford University.

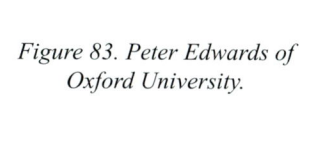

Numerous companies in the UK are also engaged in hydrogen technology research and their activity level has increased dramatically over the past few years. The knowledge and expertise of the UK industry spans the full length of the commercial value chain, from R&D to systems integration, and from finance to services. As of 2007, over 100 companies are contributing to the creation of the global fuel cell industry in the UK and they have begun to be launched on the London Stock Exchange's AIM market, recognising The City as a key source of investment for fuel cell companies. At the beginning of 2007, seven companies were listed on the AIM with combined market cap of over 500 million pounds sterling.

UK research activities on hydrogen are wide ranging from socioeconomic and public acceptance studies to atomic scale studies of solid state storage. The UK has over 35 academic and contract research groups active in fuel cells and hydrogen research, as well as a number of contract research organisations with relevant experience. The UK academic base exhibits a high degree of collaboration and maintains strong links with Germany, the USA, Canada, Japan and China. In 2003, academics published over 100 papers directly related to fuel cells and hydrogen. In the field of hydrogen production, we find universities such as Glamorgan, Northumbria and Reading. In hydrogen storage, there are groups at Rutherford Appleton, Oxford, Birmingham and Nottingham. The Applied Alloy chemistry Group at the University of Birmingham has a strong basis in rare-earth magnet research, and alongside this, the group now pursues research interests in several aspects of hydrogen related technologies. This group overseen by Rex Harris and David Book particularly focuses on material for hydrogen storage applications and hydrogen purification membranes. At present, one major project involves the conversion of a diesel engine canal boat to a clean and silent hydrogen fuel cell and an electric motor propulsion system. The boat will be used as a demonstration to the public of a hydrogen system in practice as well as a vessel from which to communicate the importance of reducing CO_2 emissions and the effects of climate change. Much work has been devoted to infrastructure for vehicle refuelling, some of which is led by David Hart from the world known Fuel Cell and Hydrogen Research Group at the Centre for Energy Policy and Technology at Imperial College London. Hydrogen energy system teams are also working at University of Manchester Institute of Science and Technology (UMIST), Bath, Loughborough and Strathclyde.

A number of impressive demonstration projects are also going on in the UK. To list a few of them, three hydrogen fuel cell buses are currently in service in London and the hydrogen fleet is to be increased to 70 vehicles by 2010 including buses. The 2012 London Olympics are intended to be a low carbon and zero waste games. Thus, low or no emission vehicles with fuel cell technology, such as fleets of cars, buses and service vehicles, will be operated and procurement for Olympics will take account of current trials of the fuel cell buses in London. In Eyemouth, Berwickshire, the nation's first Home Energy Centre was installed in September 2005. This is a cutting edge micro Combined Heat and Power unit (micro-CHP) powered by a fuel cell that runs on hydrogen converted from natural gas, and is part of a larger scale field test involving 100 houses across Europe. Finally we will mention some interesting and increasing regional activity in the UK. This includes the Fuel Cell Application Facility on Teesside, involving various demonstration activities; the Scottish Hydrogen and Fuel Cell Association which is providing a voice for the fuel cell community in Scotland and supporting projects such as PURE and last but not least the development of a hydrogen cluster in South Wales.

Among the most important hydrogen related organisations in the country are the UK Hydrogen Association, the H2Net, and the UK hydrogen research network. In the section on hydrogen storage we discuss the 'Grand Challenge' project and the fascinating approach taken by Professor Peter Edward's team in Oxfordshire.

United States, the pacemaker

We have seen in the sections on fuel cells how many American scientists and engineers led the way in the development of fuel cell technology with solid support from the space programme. In fact, NASA (National Aeronautics and Space Administration) has been a persistent sponsor of the hydrogen research related to space exploration and space launch projects, in addition to being a large user of hydrogen in its space rockets and in electricity production on board spacecraft for decades. Besides NASA, the U.S. has a long history in federal support for hydrogen and fuel cell R&D dating back to the 1970s. We will start by examining this further and then take a look at the centres of research, their interaction with industry and academia and finally comment on the network of various stakeholders.

Above all, the Department of Energy (DOE) ever since its establishment in 1977 has exerted strong leadership in supporting the research efforts related to hydrogen and fuel cells. Here are some of the key milestones initiated by the DOE. In January 2002, under the leadership of Energy Secretary Spencer Abraham, the DOE introduced the FreedomCAR Partnership as a research and development partnership between the DOE and the United States Council for Automotive Research (USCAR), an organisation for collaborative research among DaimlerChrysler Corporation, Ford Motor Company and General Motors Corporation. The goal of the FreedomCAR partnership was to develop and validate the technologies necessary to enable mass production of affordable hydrogen-fuelled fuel cell vehicles. Leading the way in Washington around this time were a number of individuals such as congressman Robert Walker, head of the House Science Committee and Robert Dixon and later David Garman, both of whom served for a long

Figure 84. Visiting delegation from Energy Committee of the US Congress. Senators Clinton and McCain with colleagues and participants from US and Iceland at the Blue Lagoon meeting facility in Iceland. Source: Icelandic New Energy.

time in leading positions within the Department of Energy and who were to become leading American hydrogen advocates.

Building on the FreedomCAR Partnership, President Bush announced the $1.2 billion Hydrogen Fuel Initiative during his State of the Union address in January 2003, so that "America can lead the world in developing clean, hydrogen-powered automobiles".

After the Hydrogen Fuel Initiative was announced, the existing FreedomCAR Partnership was expanded to the FreedomCAR and Fuel Partnership in September 2003 by bringing major energy companies together to focus on storage and delivery. The new partnership received a very positive review from the National Research Council (NRC) in 2005 and this accelerated the research efforts of the partnership. In parallel with such domestic drive, the DOE also stressed the importance of international cooperation under the recognition that collaborative hydrogen and fuel cell R&D would play a critical role in advancing the transition to a global hydrogen economy. These were the very roots of the International Partnership for the Hydrogen Economy, which was established in a ministerial meeting heralded by the American Energy Secretary in November 2003.

Looking back, a series of vision papers, roadmaps and legislative actions helped to shape the U.S. hydrogen energy policy. In November 2001, some 53 leading business executives, academics and federal and state government representatives as well as national laboratories´ executives gathered together in Washington DC for a National Hydrogen Vision meeting. This meeting, in turn, provided significant inputs for a sort of a manifesto, *Vision of America's Transition to a Hydrogen Economy – to 2030 and beyond*, that was released in February 2002. This work was soon followed in November 2002 by a National Hydrogen Roadmap which captures the discussion of a workshop under the same title held earlier the same year. Besides the stakeholder inputs, the DOE released a *Strategic Plan* in September 2003. The DOE subsequently published the *Hydrogen Posture Plan* in February 2004, providing more detailed overview regarding step by step development of the emerging hydrogen energy economy. In February 2005, the DOE published its Hydrogen, Fuel Cells and Infrastructure Technologies Programme,

Multi-Year Research, Development and Demonstration Plan, describing the planned research, development and demonstrations through 2015. Moreover, in December 2005, Energy Secretary Samuel Bodman unveiled a *Roadmap for Manufacturing R&D on the Hydrogen Economy* which identifies the manufacturing research and development challenges that must be met. In the meantime, the Energy Policy Act of 2005 (EPAct 2005) received a presidential signature in the late summer of the same year. This act is an important milestone for the U.S. energy policy as the first comprehensive energy legislation in over a decade. Most of all, the hydrogen provisions in the Act codified the President's Hydrogen Fuel Initiative and validated the timeline and built on programmes that were established and ongoing within the DOE and other government agencies.

Currently, a strong and dedicated team is leading the DOE hydrogen programme. One of the most powerful methodologies of the DOE in promoting the hydrogen technology has been the goal setting based on industrial input or performance. The key points have been to make hydrogen competitive with alternative technologies on dollar per mile basis, independent of the pathway used to produce and deliver hydrogen in the market by 2015. For example, the DOE hydrogen programme has set the target of reducing the cost of hydrogen production to USD 2.0 - 3.0 per gallon of gas equivalent (gge) at the pump by 2015 and of developing on-board hydrogen storage system achieving 9 per cent weight and USD 3 / kWh by 2015. This method provides a transparent process and the goals can be expected to change after annual reviews.

A number of demonstration projects have already been carried out nationwide and internationally. To list a few, Georgetown University in Washington DC started the Advanced Vehicle programme in 1983, with feasibility studies for fuel cell powered transit buses conducted with Los Alamos National Laboratory. The success of these studies led to the development of three Fuel Cell Test Bed Buses (TBBs) based on alkaline fuel cells in 1991. This was a part of the so-called Generation I Bus Programme and the buses were rolled out in 1994 and 1995. The reader should notice the very early timing of this ambitious work.

Subsequently, Georgetown University began to develop Generation II buses using 100kW phosphoric acid fuel cell (PAFC) and 100kW PEM fuel cell. The first Generation II buses started driving in 1998 and 2001. In 2006, Georgetown University started the Generation III phase of their programme. In many ways the pioneering spirit of the Georgetown bus programmes has been remarkable.

To complete the list of the most important moves in Washington we need to refer to a second bill, a long-term transportation funding reauthorisation bill that was signed into law by the President in August 2005. The law, called SAFETEA-LU (Safe, Accountable, Flexible, Efficient Transportation Equity Act-A Legacy for Users) authorises the federal surface transportation programmes for highways, highway safety and transit for the period 2005-2009. This law contains hydrogen provisions and supports such programmes as National Fuel Cell Bus Technology Development Programme, Clean School Bus Programme, Clean Fuels Grant Programme and Hydrogen Demonstration Programmes, it has significant effects on the hydrogen and fuel cell demonstrations in the transportation sector. For example the Federal Transit Administration (FTA) is operating the National Fuel Cell Bus Technology Development Programme to facili-

Figure 85. Arnold Schwartzenegger tanks up at the Davis hydrogen facility in California. Source: UC Davis.

tate the development of commercially viable fuel cell buses and USD 49 million will be funded over the fiscal years 2006 to 2009.

Late in 2005 the DOE and the FutureGen Industrial Alliance signed an agreement to build a prototype of the fossil-fuelled power plant of the future named "Future-Gen". The USD 1 billion government-industry project will produce electricity and hydrogen with zero emissions. In the future, the plant could become a model hydrogen-production facility to develop a new fleet of hydrogen powered cars and trucks. Other than these projects listed above, there are still many demonstration projects such as the Las Vegas energy station, learning demonstrations, hydrogen refuelling infrastructure and so on.

National laboratories and universities are also deeply involved in and contributing to the U.S. hydrogen programme. The number of laboratories is large and even larger is the number of universities attached to many of their programmes. Let us briefly look at the impressive list of the laboratories. The national laboratories across the country that are administered by DOE are strong advocates of hydrogen and fuel cell technology world wide. Many of these labs. join forces with universities and there have been formations of centres of excellence in chosen areas of the technology. Argonne National Laboratory (ANL) develops an advanced concept in nanoscale catalyst engineering that will bring PEM fuel cells for hydrogen-powered vehicles closer to the massive commercialisation. Brookhaven National Laboratory (BNL) studies an appropriate catalyst, such as titanium, that can make the hydrogen storage process suitable for practical applications. INL (Idaho National Laboratory) pursues development of and commercialisation of technologies related to production, infrastructure and utilisation of hydrogen fuel. We saw in the section on the bus trials in Washington how LANL (Los Alamos National Laboratory) was involved as it has been since the mid-1970s.

In 2004 Secretary Abraham announced that DOE was awarding the formation of three "centres of excellence" in hydrogen storage. The centres were based on the National Renewable Energy Laboratory, Los Alamos National Laboratory and Sandia Laboratory. The centres were focused on Metal Hydrides, Chemical Storage and Carbon-Based Materials respectively. Each centre had seven universities linked as well as a number of industrial partners besides other federal laboratories. To give the reader an insight into the university participation in projects (which we can not do in every case in the U.S. survey) we mention that in the Metal Hydride centre there are Stanford, CalTech, Pittsburgh, Carnegie-Mellon, Utah, Hawaii, Nevada-Reno and Illinois universities participating in this particular centre.

Lawrence Berkeley National Laboratory (LBNL) investigates new classes of materials that can efficiently story hydrogen aboard cars. LLNL (Lawrence Livermore National Laboratory) studies a hydrogen storage concept that may demonstrate an advantage over existing technologies. NETL (National Energy Technology Laboratory) conducts research and development efforts related to separating hydrogen from fossil fuels and utilizing this hydrogen for distributed and centralised power generation. NREL (National Renewable Energy Laboratory) always has a special role in hydrogen research as many want to make a clear distinction between renewable based hydrogen and fossil based one. NREL touches on most aspects in the production-to-utilisation process of hydrogen. At Oak Ridge, the ORNL (Oak Ridge National Laboratory) materials and processes are developed for fuel cell systems and for the practical generation, storage and delivery of hydrogen as an energy carrier. PNNL (Pacific Northwest National Laboratory develops technologies for cost effective production of hydrogen and has for example been focusing on gasoline reforming for PEM applications. This laboratory also leads the all important safety programme in the nation. We mentioned the important work of Sandia National Laboratories (SNL) in our discussion about storage and the famous Sandia data base earlier in the book. Last but not least we should mention SRNL, the Savannah River National Laboratory which is also engaged in solid-state hydrogen storage.

When looking at the situation in the United States it seems to me that there is no lack of brainpower to tackle hydrogen energy. This refers to the research institutions and universities. The reader may ask if all players around the market-table are involved? In a free market system, the power of the stakeholders on the market is enormous. In this context I have admired Shell and Chevron for their involvement in hydrogen – but I have always missed Exxon who have been reluctant, to say it politely, to recognise a future beyond fossil fuels. Many other corporations stand in the grey shadow of Exxon and only an awakening among the shareholders will shift this dinosaur mentality. In the positive spirit of this book I will continue waiting for a shift.

The integrated network of national laboratories, their cooperating industrial partners and universities in the United States has no match elsewhere on the Planet. Therefore the expectation level for paradigm shift in hydrogen research is very high when these hubs of knowledge are considered and evaluated and the market situation is right.

The United States is a trendsetter in the world of hydrogen energy and we may ask where the stakeholders of the hydrogen energy society join their forces on a more visionary and non-governmental level. The US National Hydrogen Association is a membership organisation founded by a group of ten industry, university, research, and small business members in 1989. Today, the NHA's membership has grown to over 100 members, including representatives from the automobile industry; the fuel cell industry; aerospace; federal, state, and local government; energy providers; and many other industry stakeholders. The NHA under the steadfast presidency of Jeff Serfass serves as a catalyst for information exchange and cooperative projects and provides the setting for mutual support among industry, government, and research/academic organisations. The annual meetings of the NHA are among the highlights of the world hydrogen events with attendance from all over the globe.

I devoted this chapter to the pacemakers. The Americans put the first man on the Moon. We live in a market driven world which is constantly threatened by external conditions and needs very clear goals and leadership to attain those goals. Sometimes I wonder that the huge initiative taken in Washington at the beginning of the new millennium could have developed into a "space-race-like" action if more money had been devoted to this good cause. The U.S. is undoubtedly the pacemaker. Over the past years the DOE has been allocating up to in excess of 400 million dollars to hydrogen annually. A space-race style devotion would perhaps call for ten times larger sums and this is still possible. Also, in retrospect, it needs to be kept in mind that the political turmoil felt in the wake of the 9-11 events, weakened the focus somewhat.

My hope is that a new presidency will rekindle the old flame started in the first years of the millennium in Washington. The pacemaker needs to be in his best form for completing the run and I cannot but admire the good work done so far by those Americans dedicated to the development of the new fuel.

PART VI
NEAR A JOURNEY'S END

Dear Reader,

In our passage we have covered the scene in time and space; gazed into the future, then back to the origins of our universe, as well as obtaining an insight into the harnessing of energy with emphasis on renewable sources and carriers of energy. Furthermore, we have come to know hydrogen and fuel cells, the storage and distribution of the lightest element and finally undertaken a trip around the world to learn about the most recent developments in hydrogen technology world-wide.

In the process of creating and writing this book, which has taken four years, considerable development has taken place in the field of environmental awareness. This in turn has resulted in appreciable silencing of the voices of complete doubt that did characterise the discussion a few years ago.

Increased scientific evidence has piled up and now it is generally recognized that besides the natural causes of climate change – there is an anthropogenic influence of notable dimensions. For me as a writer it has been wonderful to witness this global change of attitude during the writing of the book.

During our journey together we have reached an understanding of the finiteness of fossil energy resources. At the same time we have studied and accepted that this energy still remains out there and will not disappear quickly – humankind will undoubtedly do its utmost to squeeze the last drops of oil out of the ground. The coal reserves of the world are also far from used up, they can last for centuries and we expect methods of burning coal to be replaced by methods providing higher purity.

At the same time we have examined the possibilities opened up by various renewable energy sources as well as nuclear energy in the role of replacing fossil resources. However, as long as the narrow economical criteria do not take into account various external costs related to environmental problems, public health etc., renewable energy will run into serious difficulties in competing with fossil energy. This is in fact the core of the problem.

My feeling right now is that there are strong signs of changes in the understanding of the ecological/economical forces behind it all. This will lead to more inclusion of externalities; there will be increased urge to create a tax on carbon dioxide and other green house gases. Again, this will be reflected in increased quality of energy utilization in general and an increased emphasis on renewables.

When considering fuels, we have looked at various ways to reduce green house gas emissions which are related to the fuel use. We have considered bio-fuels, which enjoy enormous popularity today, to replace fossil fuels – and noticed that in the life-cycle of these fuels in many countries there is still a high demand for the use of fossil fuels.

From these considerations we investigated increased efficiency of energy use, for example by looking at the utilization of the braking-power of cars in the hybrids. In our analysis of the car of the future, the new electric car, we argued that a fuel cell would greatly enhance the range of the fuel and be superior to the good-old combustion engine drive train. How to store energy for the electric car of the future has been left as a key question.

I have used the term "taming of the proton" about the technological task to harness hydrogen as an energy carrier. The associated challenges of this exciting technology are enormous and present fundamental physical as well as chemical difficulties. Hydrogen is even more challenging to tame that Shakespeare's Shrew! An for me as a scientist I have taken great pride and joy in this worthy endeavour. The strong feeling of participating in a paradigm shift from the Carnot era to the post Carnot times of Gibbs free energy makes our times so exciting.

Our voyage has taught us how gradual successes in hydrogen technology are finally making it competitive when compared with the good old traditional energy infrastructure. The new technology demands electric motors and fuel cells. The electricity can be stored in batteries, but for large quantities we need to store the energy in the form of physical fuel. Our ideal physical fuel contains a very large proportion of protons indeed. Our insight into the history of energy conversion has given us insight into various victories on the road to success.

The future will reveal how humankind manages to resolve the problem we are facing. We will witness solutions that gradually move the world closer to decarbonisation and cleaner energy sources and carriers. Even during the next years I expect that we will be presented with many attempts to solve the present challenge – long-term as well as short-term solutions. The final solution may reveal itself to our children, perhaps even our grandchildren around and after the middle of the century.

In the ever lasting interaction of humankind and nature, the taming of the proton has not been completed. It is indeed among the greatest challenges of the human spirit ever witnessed.

Thank you for your company.

REFERENCES & FURTHER READING

References and Further Reading

Section I.

Verne, Jules.(1874). "L´ille Mysterieuse", *Hetzel*, Paris English Translation published by Signet Classic 1986.

Section II.

Ausubel, J.H.(1996). "Can Technology Spare the Earth?" *American Scientist* 84(2): pp 166-178.

Brown L.R.(2006). "Plan B 2.0, Rescuing a Planet Under Stress and a Civilization in Trouble". *Norton*.

Brown, LeMay and Bursten.(2006). "Chemistry: The Central Science", *Pearson and Prentice Hall*, New Jersey.

Deng, Yenke.(2005). "Ancient Chinese Inventions", *China Intercontinental Press*.

Epstein, Paul. "Climate Change and Human Health" *New England Journal of Medicine*, pp 1433-1436. Oct. 6th 2005.

European Energy and Transport – Trends to 2030. Office for Official Publications of the European Communities, Luxembourg 2003.

Flannery, T.(2005). "The Weathermakers", *Grove Press*.

Gipe, P.(2004). "Wind Power", *Chelsea Green Publishing Company*.

Hawking, S.(1988). "A Brief History of Time – from the Big Bang to Black Holes", *Bantam Books*.

Hoffmann, P.(2001). "Tomorrows Energy", *MIT Press*.

"HyWays: a European Roadmap, Assumptions and robust results form phase I, U".

Buenger.(2005). "private communications", 2005/6.

IEA / Solar PACES, News issues 2001-2003.

"International Energy Outlook 2005 Highlights".(2005), IEO.

IPCC Special Report on Carbon Dioxide Capture and Storage, Summary for Policymakers, the 8th session of IPCC Working Group III, September 2005.

IPHE Scoping Papers ILC-037-05, Washington, March 1, 2005

Jakobsen, V. E., Hauge F., Holm, M. and Kristiansen, B.(2005). "Bellona report", Oslo.

Klare M.T.(2001). "Resource Wars", *Henry Holt Publishing*.

Lane, Nick.(2002). "Oxygen, The Molecule that made the World", *Oxford University Press*.

Mills, E.(2005). "Insurance in a Climate of Change", *Science,* 309, pp 1040-1044.

Pakala, S. and Sokolow, R.(2004). "Stabilization Wedges: Solving the Climate Problem for the Next 50 Years with Current Technologies", *Science*, Vol. 305.

"Renewable Energy World".(2005). Vol. 8, 5..

"Renewables for Power Generation Status and Prospects" *International Energy Agency and OECD*, Paris, 2003.

Ridgen, John S.(2002). "Hydrogen, The Essential Element", *Harvard University Press*.

Rifkin, J.(2002). "The Hydrogen Economy: The Creation of the Worldwide Energy Web and the Redistribution of Power on Earth", *Jeremy P. Tarcher/Putnam*.

Riis, Tryggve "Private communications".

Saad, M. A.(1997). "Thermodynamics, Principles and Practise", *Prentice Hall*.

Saunders, M.A.(1999). "Earth´s future climate, Philosophical Transactions, Royal Society Millenium issue", Series A, Volume 357.

Sigfusson, T.(2007). "Pathways to Hydrogen as an Energy Carrier", Philosophical Transactions of the Royal Society A. Volume 365, Number 1853 / April 1.

Smil, V.(2003). " Energy at the Crossroads, Global Perspectives and Uncertainties", *MIT Press*.

Sturluson, Snorri.(1996) "The Poetic Edda", Oxford World´s Classics. Carolyne Larrington /Ed.), *Oxford University Press*.

Wald, M.L.(2007). "Is ethanol for the long haul?" *Scientific American*, January issue p.42-49.

Section III.

Abrahamsson, M., Berglund Baudin, H., Tran, A., Philouze, C., Berg, K. E., Raymond-Johansson, M. K., Sun, L., Åkermark, B., Styring, S. and Hammarström, L. Ruthenium-Manganese.(2002). "Complexes for Artificial Photosynthesis: Factors Controlling Intramolecular Electron Transfer and Excited State Quenching Reactions". *Inorg. Chem.* **41**, 1524-1544.

References & Further Reading

Winter, Carl-Jochen (Ed.).(2000). "The Energies of Change – The Hydrogen Solution" *Gerling Akademie Verlag*.

Ceperley, D. and B. Alder.(1980). Phys. Rev. Lett. **45**, 566.

Eichlseder, H. (Ed.).(2006). "Proceedings of the First International Symposium on Hydrogen Internal Combustion Engines", *Graz University of Technology*, Austria.

Fujishima, AK and Honda, K.(1972). "Electrochemical photolysis of water in a semiconductor electrode", *Nature (London)*, 238: pp 37-38.

Heben, Michael, Zhang, Shengbai et al.(2005). "Phys". Rev. Lett. 94, 155504.

Heinzel, A, Mahlendorf, F. and Roes, J.(2006). "Brennstoffzellen", *C.F.Mueller Verlag*, Heidelberg.

IPHE Scoping Papers ILC-037-05, Washington, March 1, 2005

Jain I.P. and Abu Dakka M.I.S., (2002) "Hydrogen absorption-desorption isotherms of $Na_{(28.9)}Ni_{(67.55)}Si_{(3.55)}$" *International Journal of Hydrogen Energy*, pp 27 395-401.

Jonsson, Hannes, "Private communication".

Jörissen L., Gogel V., Kerres J. and Garche J., J.(2002). "Power Sources", pp 105, 267.

Kapischke, J, Hapke, J.(1998). "Measurement of pressure-composition isotherms of high-temperature and low-temperature metal hydrides". *Experimental Thermal and Fluid Science*, pp 18 30-81

Kitamura, H. and Ichimaru, S.(1998). J. Phys. Soc. Japan 67 (3), 950.

Koppel, T.(1999). "Powering the Future, The Ballard Fuel Cell and the Race to Change the World". *John Wiley and Sons*, Canada.

Larminie, J. and Dicks, A.(2002). "Fuel Cell Systems Explained", *John Wiley and Sons*, England, 5th reprint.

Luzzi, A. Bonadio, L. and McCann, M. (Ed.).(2004). "In Pursuit of the Future, Hydrogen Implementing Agreement", *IEA*, 2004.

Magro, W. R. Ceperley, D. M. Pierleoni, C and Bernu, B.(1996). Phys. Rev. Lett. 76, 1240.

Majzoub, Eric H. et al.(2005). "Phys". Rev. B Volume 71, 024118.

Nitsch, Joachim and Winter, Carl-Jochen.(1988). "Hydrogen as an Energy Carrier, Technologies, Systems and Economy". *Springer Verlag*.

Oslund, H.G. and Alexander, J.(1963). "Oxidation rate of sulfide in sea water, a preliminary study". *J. Geophys. Res.*, 68(13): 3995.

Ostwald, W.(1984). "Z Electrochemie", 1, 122.

P. Dantzer, M. Pons, A. Guillot.(1994). "Thermodynamic properties in the non-equilibrium $LaNi_5-H_2$ System". *Zeitschrift für Physikalische Chemie*, pp 183 205-212

Paul, Anderson and Edwards, Peter.(2005). "Chem". *Commun.* pp 2823.

Riis, Tryggve "Private communications".

Sandrock, G.(1999). "A panoramic overview of hydrogen storage alloys from a gas reaction point of view". *Journal of Alloys and Compounds*, pp 293-295 877-888

Sigfusson, T.I. Wang, S. and Arnason, B.(2005). "Hydrogen Production and Utilization from Geothermal Gasses", Proceedings of The International Hydrogen Energy Congress and Exhibition, Istanbul 13-15 July, 13 p.

Soerensen, Bent.(2005). "Hydrogen and Fuel Cells: Emerging technologies and applications", *Elsevier*.

Swickardi M. and Bogdanovic B.(1997). "J. Alloys Compounds", vol 1, p.253.

Uchida, H, Sato, M, Moriwaki, O.(1997). "Hydrogen absorption and desorption isotherms in the solid solution regions of the $LaNi_5$-H system". *Journal of Alloys and Compounds* pp 253-254.

Vielstich, W., Lamm, A., and Gasteiger, H.A.(2003). "Handbook of Fuel Cells, Vol.1, Fundamentals, Technology and Applications", *John Wiley and Sons Ltd. Chichester*.

Weir, S.T., Mitchell, A.S. and Nellis, W.J.(1996). "Metallization of Fluid Molecular Hydrogen", Physical Review Letters 76, 1860.

Westphal, C.(1880). "Apparat zur Erzeugung electrischer Ströme" *DRP* 22393.

White, Guy, Kendall.(1968). "Experimental Techniques in Low-Temperature Physics", *Oxford University Press*.

Yukawa, H., Nakatsuka, K., Morinaga, M.(2000). "Design of hydrogen storage alloys in view of chemical bond between atoms". *Solar Energy Materials & Solar Cells*, pp 62 75-80.

Zhao Y.,. Kim Y-H., Dillon A.C., Heben M.J. and Zhang S.B.(2005). "Hydrogen Storage in Novel Organometallic" *Buckyballs. Phys. Rev.* Lett. 94, 155504.

Zhitao Xiong, Chan Keong Young, Guotao Wu, Ping Chen, Wendy Shaw, Abi Karkamkar, Thomas Autrey, Martin Owen Jones, Simon R. Johnson, Peter P. Edwards, William I.F. David (2008). Nature Materials, 7, 138-141.

Section IV.

Andersen, P.D., Joergensen, Holst, Eerola, B., Kojonen, A., Loikkanen T and Eriksson E.A.(2005). "Building the Nordic Research and Innovation Area in Hydrogen", Nordic Energy Research Summary Report, January. ISBN 87-550-3401-2; ISBN 87- 550-3402-0 (internet).

Ausubel, J.H., Marchetti, C. and Meyer, P.(1998). "European Review", Vol. 6, No 2. 137-156.

"European Energy and Transport – Trends to 2030". *Office for Official Publications of the European Communities*, Luxembourg 2003.

References & Further Reading

"HyWays: a European Roadmap, Assumptions and robust results form phase I, U".

Buenger.(2005). "Private Communications", 2005/6.

Karlsson, B. and Quintiere, J.G.(2000). "Enclosure Fire Dynamics", *CRC Press*.

Kurani, K., Turrentine, T., Heffner, R. and Congleton, C.(2004). "Prospecting the Future for Hydrogen Fuel Cell Vehicle Markets," chapter in D. Sperling and J. Cannon, eds., *The Hydrogen Energy Transition*, Elsevier Press.

Ogden, J.(2004). "Where Will the Hydrogen Come From? Systems Considerations and Hydrogen Supply," chapter in D. Sperling and J. Cannon, eds., *The Hydrogen Energy Transition*, Elsevier Press, 2004.

Ogden, Joan; Yang, Christopher; Johnson, Nils; Ni, Jason; Lin, Zhenhong.(2005). "Technical and Economic Assessment of Transition Strategies Toward Widespread Use of Hydrogen as an Energy Carrier" Hydrogen Pathways Report to the United States Department of Energy. January. UCD-ITS-RR-05-13.

Owen, A.D.(2004). "Environmental externalities, market distortions and the economics of renewable energy technologies", *The Energy Journal,* vol. 25 pp 127-156; and references citet therein.

Section V.

Árnason, B. and Sigfusson, T.I.(2000) "Iceland - A Future Hydrogen Economy". *International Journal of Hydrogen Energy*, Vol. 25, No. 5, May, pp. 389-394.

"BP Statistical Review of World Energy 2006", BP, June 2006.

Chopra, S.K.(2005). "Towards Hydrogen Energy Economy in India" *presented at the UNU Conference,* , Ministry of Non-Conventional Energy Sources, India, November.

"Deployment Strategy", European Hydrogen & Fuel Cell Technology Platform, August 2005.

"Energy in Australia 2006", Australian Government Department of Industry Tourism and Resources, Australia, 2006.

"Energy Research Roadmap", Ministry of Research Science and Technology, New Zealand, December 2006.

"Fuel Cell Commercialization Roadmap", Industry Canada, March 2003

"Hydrogen: A new possible bridge between mobility and distributed generation (CHP)" *presented at the 19th World Energy Congress*, Valentino Romeri, Rasfin Sim, Italy, September 2004.

IPHE Country Updates *presented at the IPHE Steering Committee Meetings and Implementation-Liaison Committee Meetings*, IPHE website (www.iphe.net).

"National Evaluation System for Public R&D Programs in China*" presented at the KISTEP / WERN International Workshop in Korea*, Yan Fang, China National Center for Science and Technology Evaluation, China, May 2005.

"Norwegian Hydrogen Strategy", Ministry of Transport and Communications & Ministry of Petroleum and Energy, August 2005.

"Powering Our Future: Draft New Zealand Energy Strategy to 2050", Ministry of Economic Development, New Zealand, December 2006.

"Projected Development of Fuel Cells and Hydrogen Technologies in Japan", British Embassy Tokyo, June 2006.

"Research on Hydrogen and Fuel Cells in Italy" *presented at the 6th General Assembly of HyWays,* Antonio Mattucci, ENEA, Italy, October 2006.

Sigfusson, T.I., "L´ile de Jules Verne", Chapter in *Decouverte*, Revue du Palais de la decouverte, Paris, pp 64-73.

Sigfusson, Thorsteinn I.(2005). " Renewable energy Island", Renewable Energy 2005, Flin David (Ed.), Offical publication of the World Renewable Energy Network and UNESCO, pp95-97, 2005.

Sigfusson, Thorsteinn I.(2005). "Transport Fuels from Domestic Sources: The Icelandic Hydrogen Project". In Technology in Society, (Ed.) Ö.D.Jónsson and E.H. Huijbens, *University of Iceland Press* pp 251-266.

"Status of Research and Development Activities in Italy" presented at the 2nd network committee meeting of HY-CO, ENEA, Italy, March 2005.

"Strategic Overview", European Hydrogen & Fuel Cell Technology Platform, June 2005.

"Strategic Research Agenda", European Hydrogen & Fuel Cell Technology Platform, July 2005.

"Towards a National Hydrogen & Fuel Cell Strategy", Industry Canada, 2006.

"UK Fuel Cell Development and Deployment Roadmap", Fuel Cells UK, 2005.

GLOSSARY

ESSENTIAL HYDROGEN GLOSSARY

AFC – Alkaline fuel cell. Can only be operated with pure hydrogen and air where CO_2 has been removed. Operating temperature usually 60 to 90°C.

Biomass – Plant or vegetation or agricultural waste useable for fuel or as an energy source such as wood chips, bales of straw, liquid manure, organic wastes etc.

Catalyst – A catalyst is a material that facilitates, accelerates etc. a chemical reaction retaining its own properties and without being consumed.

CCS – Carbon capture and storage. Refers to the capture of carbon contained in emitted CO_2.

CGS – Compressed gas storage device for gases (e.g. hydrogen or natural gas) at room temperature under high pressure.

Compressor – Device for increasing gas pressure or gas flow rate.

Cryogenic – Greek krýos: cold, frost. Applied to gases cryogenic refers to low temperatures where the gases are in their liquid phase. For natural gas the boiling temperature (where the phase transition from liquid to gaseous occurs) is -161.5°C (111.5 K) and for hydrogen it is -253°C (20 K).

Deuterium – An isotope of hydrogen where the nucleus consists of one proton and one neutron.

Density of liquid hydrogen – 70.8 kg/m^3

DMFC – Direct Methanol Fuel Cell.

DoE – The Department of Energy, United States.

Efficiency – The efficiency of a heat engine relates how much useful power is output for a given amount of heat energy input

Electrolyser – A device for using electric current to split water. In the case of water it is converted to hydrogen and oxygen. Electrolysis can be used in various other processes such as for example in the aluminium industry.

Energy carrier – Medium (gaseous, e.g. natural gas, hydrogen; liquid, e.g. petrol, biofuels; solid, e.g. wood, coal) in which energy is stored in chemical form; by means of energy carriers. Energy is storable and transportable. Non-material energy carriers are e.g. electricity and solar radiation. Within certain limits and with certain losses energy carriers can be converted into one another (e.g. solar radiation into electricity, electricity into hydrogen, hydrogen into electricity, electricity into light etc.).

Fuel Cell – A fuel cell is a static electrochemical device (no moving parts) that converts the chemical energy of a fuel, such as hydrogen, and an oxidant, such as oxygen, directly to electricity and heat.

GGE – Gallon Gasoline Equivalent (GGE)- is the amount of alternative fuel it takes to equal the energy content of one liquid gallon of gasoline. The term is very often used by for example the Department of Energy in the US.

Heat engine– A physical or theoretical device that converts thermal energy to mechanical output. The mechanical output is called work, and the thermal energy input is called heat.

H_2 – The symbol for the hydrogen molecule composed of two hydrogen atoms.

Heating value. HHV and LHV – Energy content of an energy carrier. Higher and lower heating value are distinguished. Higher heating value: total energy content of the energy carrier. Lower heating value: energy content reduced by the condensation energy (latent heat) of the product gas (the steam in the product gas, to be exact).

Hydrogen – H is the chemical symbol for hydrogen, the lightest element of the table of elements and the most abundant element of the universe. In general, hydrogen will be found in molecular form, i.e. as a hydrogen molecule composed of two hydrogen atoms (H_2), or in other compounds (e.g. in water – H_2O, organic substances). Hydrogen as secondary energy carrier is seen as the key component of a global renewable world energy supply.

Hybrid Electric Vehicle – A vehicle combining a battery-powered electric motor with a traditional internal combustion engine. The vehicle can run on either the battery or the engine or both simultaneously, depending on the performance objectives for the vehicle. FC Hybrid electric vehicles use hydrogen and fuel cell to provide power and extend range.

Hydrogen economy – Energy economy where hydrogen is used as a secondary energy carrier.

Hydrogen liquefaction – liquefaction of hydrogen, which is gaseous at room temperature, by cooling it below -253°C (20 K).

IAHE – International Hydrogen Energy Association, established 1973.

ICE – Internal Combustion Engine. An engine that converts the energy contained in a fuel inside the engine into motion by combusting the fuel. Combustion engines use the pressure created by the expansion of combustion product gases to do mechanical work.

IEA HIA – International Energy Agency – Hydrogen Implementation Agreement.

IPHE – International Partnership for the Hydrogen Economy, established in Washington 2003.

Isotope – One of several forms of an element having the same atomic number but differing atomic masses. Hydrogen has three isotopes, protium, deuterium and tritium.

Kelvin temperature scale – bears the name of its inventor Lord Kelvin. The Kelvin scale defines an absolute zero temperature 0 K, which corresponds to -273 degrees Celsius. In this way to convert Kelvin degrees to Celsius, one has to add 273 degrees to the Kelvin figure.

LH2 or LH$_2$ – Liquid hydrogen

MCFC – Molten carbonate fuel cell; with molten alkaline carbonate electrolyte; operating temperature 600 to 650°C; fuel: carbon containing gases (e.g. natural gas, synthesis gas).

Metal hydride storage – Device that can store hydrogen by use of a metal alloy. The hydrogen is soaked into the alloy like into a sponge, binds chemically and fills the spaces in the crystal lattice of the alloy

Ortho- and Para-Hydrogen – Two types of hydrogen molecules where the nuclei of the hydrogen atoms have two different types of proton-spin. The two types have a resulting very different latent heat which affects the energy needed to liquefy hydrogen.

PAFC – Phosphoric acid fuel cell. Contains phosphorous electrolyte and uses pure hydrogen as a fuel. The working temperature is around 200°C.

Pascal – Unit of pressure. Mega Pascals (SI pressure unit); one MPa corresponds to a pressure of 10 atmospheres (10 bar$_{abs}$).

Partial oxidation – Conversion of hydrocarbons (diesel, residual oil etc.) into a synthesis gas that consists of hydrogen, carbon monoxide (CO) and carbon dioxide (CO_2). The necessary energy is supplied by the combustion („oxidation") of parts („partial") of the feedstock in the process itself. Partial oxidation is a common process for the production of hydrogen (the synthesis gas is converted into pure hydrogen by converting the carbon monoxide and water into carbon dioxide and hydrogen and by subsequently separating the carbon dioxide).

PEFC or PEMFC – Proton exchange membrane fuel cell. Uses a solid electrolyte that can conduct protons and uses pure hydrogen as a fuel. Operates normally at 60-80 °C.

Photobiological water splitting – There are different biological processes that liberate hydrogen or where hydrogen is produced as an intermediate product. Photobiological processes as e.g. photosynthesis use the solar radiation as source of energy, while fermentation processes that take place in the absence of light take advantage of the energy stored in the feedstock (e.g. glucose). There are several first efforts to use photobiological water splitting for the technical production of hydrogen.

Protium – The simplest of hydrogen atoms carrying one electron surrounding a proton as nucleus.

Proton – The subatomic particle carrying positive charge – the nucleus of the simplest of hydrogen atoms, the protium.

Renewable energy – Form of energy which is never exhausted because it is renewed by nature (within short time scales; e.g. wind, solar radiation, hydro power). Geothermal energy, although requiring a different timescales is also counted as renewable.

Renewables – Renewable energy sources.

SOFC – Solid oxide fuel cell. A fuel cell with oxygen ion conducting ceramic electrolyte; operating temperature 800 to 1000°C; fuel: pure hydrogen, carbon containing gases (e.g. natural gas, synthesis gas).

Steam reformer – A device for steam reforming.

Steam reforming – Catalytic conversion of light hydrocarbons (biomass, fossil energy carriers e.g. natural gas) producing a synthesis gas that consists of hydrogen (H_2), carbon monoxide (CO) and methane (CH_4). The process is heat consuming. Steam reforming of natural gas is a common process for the production of hydrogen (the synthesis gas is converted into pure hydrogen by converting the carbon monoxide and water into carbon dioxide and hydrogen and by subsequently separating the carbon dioxide).

Tritium – The unstable isotope of hydrogen where the nucleus consists of one proton and two neutrons.

Units of energy – This book uses the Joule (J) as a main unit. A Watt is a measure of the power of one Joule of energy being used in a second. The prefixes kilo, Mega and Giga are used for one thousand, one million and one billion respectively.

We sometimes use the British Thermal Unit Btu for historical reasons. One Btu equals 1/180 of the heat required to raise the temperature of one pound (1lb.) of water from 32 F to 212 F at a constant atmospheric pressure. It is about equal to the quantity of heat required to raise one pound (1 lb.) of water 1 F.

The largest unit of energy we use in the book is Quad. It stands for a quadrillion Btu's. The annual energy use of humankind is about 440 Quads. A Quad is about a billion billion, 10^{18} Joules.

For fossil fuels we often use the measure barrel. Barrel of oil equivalent (boe) = approx. 6.1 GJ (5.8 million Btu), equivalent to 1,700 kWh. "Petroleum barrel" is a liquid measure equal to 42 U.S. gallons (35 Imperial gallons or 159 litres); about 7.2 barrels oil are equivalent to one tonne of oil (metric) = 42-45 GJ.

Van Allen Radiation Belt – A torus of energetic charged particles in the form of plasma around Earth, held in place by Earth's magnetic field. The Van Allen Belts are closely related to polar aurora where particles strike the upper atmosphere and form the Northern and Southern Lights.

Well-to-Wheel Efficiency – The efficiency of a given energy carrier from its well to the use to power the wheels of an automobile.

INDEX

Index

A

Aabo Academi 176
Abdel-Aal, Hussein 131
Abdul Kalam, A.P.J., 167
Abraham, Spencer 135
Abruzzo 168
Accumulated Cyclone Energy (ACE) 19
acetic acid 58
activation barrier, 78
Aegean islands 37
Aeolian islands 37
Aerostats 177
AFC 203
Agder College 173
Agency for Environment and Energy Management 153
agricultural
    crops 57
    soil management 43
    waste 33
agro-industrial waste 57
Aichi, expo in 170
AIM market, London Stock Exchange's 181
air conditioning system 5
Air Liquide 153-4, 169
aircraft, TU 155, 177
Akira Mitsui Award 133
alanates 80
algae, green 12
alkaline 91
    fuel cells 97
Alleau, Thierry 152
Allis-Chalmers 88
allothermal 64
Alps 4
aluminium smelters 158
ammonia borane 82
Amsterdam 109
anaerobic
    conditions 60
    digestion 35
Anatolia, Western 37

Anderson, Paul 81
Andreassen, Knut 173
Andresen, Arne F. 173
Angstrom
    Laboratory 177
    Power 146
animal manure 57
Annapolis Royal station 38
anode 63, 91
Ansaldo Fuel Cells 169
Antarctica 141, 142
Apollo programme 90
aramide fibre 73
Arcotronics Fuel Cells 169
Arezzo 168
Argonne National Laboratory (ANL) 185
Aristotle 49
Arnarson, Ingolfur 160
Arnason, Bragi 157, 159-61, 164-5
Arnason, Hjalmar 161-2, 164
Ásgrímsson, Halldór 165
Association Lorraine for the Promotion of Hydrogen 152
Aston, F.W. 50
Athens 14
Atlantic Ocean 4, 64, 158
Atomic Energy Commission 153
Aurora Borealis 9
Auroras 10
austenitic steel 129
Australia 27, 39, 109, 141
Australian National University 142
auto-thermal reforming 56
autoignition temperature 128
automobile 22
    applications 103
Autrey 82
Axane 154
Azerbaijan 15
Azores 37

B

B-Big business is Back. 120

Bacon 88
    fuel cell stack 88
bacteria, thermophilic anaerobic 58
Baden-Württemberg 122, 154
Baikonur cosmodrome 177
Baku 15
Balkans 21
ball-milling 81
Ballard, Geoffrey 90, 147
Ballard Power Systems 110, 146
Barcelona 109
barium cerate perovskites 58
Bartholomy, Obadiah 113
Bath, University of 181
batteries, laptop computer 95
battery 91
    nickel metal hydride (NMH) 79-80
Bauer, Emil 90
Bavaria 154
Bay of
    Biscay 39
    Fundy 38
BAYSOLAR 154
Be(BH$_4$)$_2$ 81
Beck, Nick 134
Becquerel, Edmund 29
Beijing 133, 148, 149
    Fu Yuan Pioneer New Energy Material Company 149
    Hydrogen Park 150
    Lu Neng Power Sources Company 149
    Olympic Games 150
Bell Laboratories 29
Bellona 43
Benoit Fourneyron 32
Berwickshire 182
beryllium 81
Bicocca Project 168
Big Bang 11
binary
    alloys 79
    systems 79
bio ethanol 144

bio resources, hydrogen from 59
bioenergy 5
biofuel(s) 33, 42
    generation 42
Biohydrogen 134
biological 57
    methods of producing hydrogen 58
biomass 33, 57, 203
    gasifiers 119
    hydrogen from 57
biomimetic
    concept 68
    hydrogen 59
    methods 4
    power plant 5
    system 5
biosphere 12
bipolar plates 96
Birmingham, University of 181
Biscay, Bay of 39
bitumen 14
Björgvin Hjörvarsson 177
Black liquor 58
Black Sea 70
BMW 85, 156
boats 5
Bob Zweig 132
Böblingen 98
Bockris, John 132
Bodman, Samuel 184
Bogdanovic, B. 80
Bohr, Niels 50
Bonn 160
Book, David 181
borazine 82
Borgarfjordur 9
borohydrides 81
boron 27
bow shock 10
Boyle, Robert 50
BP 110, 142
Bragastofa 165
Bragi Arnason 157, 159-61, 164-5
Brazil 34, 144
Brazilian
    Association for Technical Standards 145
    Hydrogen Roadmap 144
Brinner, Andreas 154
Brisbane 133, 142
Britain 16
British Columbia 147
British Petroleum 43
Broecker, Wallace 44
bromine 60
    /Calcium/Iron process 60, 65
Brookhaven National Laboratory (BNL) 185
Broome 141
Brush, Charles F. 31
buckyballs 77
Buenos Aires 133
buoyancy 127
Burke, Andrew 113

burning 22
buses, Generation II 184
Bush, President 183
Busquin, Philippe, 150

C

cadmium 27
Calcium 60
California 35
    Air Resources Board 110
    Energy Commission 110
    Fuel Cell Partnership 110
CalTech 186
Cambridge 28
Campbell, Graham, 138
Campidano Graben 37
Campinas, State University of 145
Canada 21, 27, 90, 138, 146, 154, 181
Canadian Transportation Fuel Cell Alliance 146
Canaries, The 37
carbon 12, 41
    capture 108
    contribution 42
    dioxide
        Capture and Storage (CCS) 43
        collectors 4, 44
        nanostructures 167
        Sequestration
        Leadership Forum (CSLF) 44
    plants 119
carbondioxide emissions 158, 160
Carnegie-Mellon University 186
Carnot 88, 92
    heat engine 92
Carnot, Sadi 23
Carnot´s maximum efficiency 24
Casablanca 3, 5
Casaccia 173
Catalyst 203
cathode 63, 91
Cavendish, Henry 50, 180
Cavendish Laboratory 28
CCS 203
Cecil, W. 85
Cenex 180
CGS 203
Chadwick, J. 50
Channel. the 4
Charles, Alexandre Cesar 50
Chelles 153
chemical processes, hydrogen from 54
chemisorption 76, 78
Chevron 186
ChevronTexaco 110
chicken-and-egg problem 123
China 16, 21, 39, 141, 148-9, 181
chlor-alkali 62
chlorofluorocarbons 17
Chlorophyll 67

artificial 5, 68
Chopra, S.K. 166
CHRISGAS 57
Chrisgas 177
Christchurch 144
Christensen, Claus Hviid 176
Churchill, Winston 16
Citaro 94
clathrates 82
Claus Hviid Christensen 176
Clean Energy Research Institute 132
Clean Fuels Grant Programme 184
Clean School Bus Programme 184
Clean Urban Transport for Europe 142, 163
Climate Change, Intergovernmental Panel on 22
Clinton, Hillary, 183
clostridias 58
CNR (National Research Council) 168
Co-firing 34
CO_2 atmospheric concentrations 18
CO_2 collectors 44
Coal 25
coal 19, 157
    conversion from 42
Cocoa Beach 133
Codes and Standards 130
coke 26, 157
cold fusion 28
Columbia University 44
combustion energy 52
comets 12
Commonwealth Scientific and Industrial Research Or 142
complex
    hydrides 80
    metal hydrides 81
compressor 203
concentrated solar power systems 30
concentrating solar power 30
Connecticut Academy of Arts and Sciences 86
converting 22
Copenhagen 4, 176
corn 34
Coronal Mass Ejections 10
corrosion 109
Cotzen, J.P. 134
cracking 66
Crown Research Institutes 143
Cryogenic 203
cryojet 3
crystal lattice 39, 76
Curtin University of Technology 142
CUTE (Clean Urban Transport for Europe) 109, 174
    /ECTOS 93
    project 110

D

D1 Oils 34

Index

Daejeon 171
Daimler 156
Daimler Benz 160-2
DaimlerChrysler 34, 110, 117, 142, 149, 170
  Corporation 182
Dalian Institute of Chemistry and physics 149
Dalton, John 50
Danish
  government 131
  Hydrogen Association 176
  Space Research Agency 21
  Technical University 176
David, Bill 83
David Oddsson 165
Davis, Gray 114
Davis, University of California at 110
Davtyan, O. 90
Davy, Humphry 50
DC cable bundles 4
decarbonisation 40
decarbonisation 25, 129
  of power plants 42
deep drilling 36
deflagration 52
  -to-detonation 128
Deisenhofer 60
Democritus 50
Denmark 4, 31, 119, 173
density of liquid hydrogen 203
Department of Energy (DOE) 182
Department of Trade and Industry (DTI) 180
desalination plant 5
Det Norske Veritas (DNV) 176
Deuterium 51, 203
Dewar, James 73
Dicks, Andrew 104, 142
diesel 42
diffusion coefficient in air 127
diffusivity 127
dimethylether 58
Diogenes Laertus 49
direct
  fuel cells 92
  methanol fuel cell 101, 102
dissociated components 57
Dixon, Robert 135, 182
DLR German Aerospace Centre 154
DMFC 203
DoE 203
Dönitz, Wolfgang 154
Drake, Edwin L. 14
Draugen 43
driftwood 157
Dunkerque 153
DuPont Corporation 90
dynamo, electric 23
Dynetek 72, 148

E

E-Energy Entrepreneurs and Smart

Policies 120
Earth 4-5, 10, 17
  surface temperature 20
  magnetic field 10
East Africa 38
Eastern Europe 115
Ebara-Ballard 169
economics 122
economy, hydrogen energy 5
ECTOS 163, 174
  project 110
EDB/Elsam 100
Edinburgh University 50
Edison 169
education 121, 172
  hydrogen 121
Edwards, Peter 81, 83, 182
efficiency 17, 203
Eggert, Anthony 114
Egypt 30, 131
Einstein, Albert 27, 122
Eisenhower, President 124
electric dynamo 23
Electricité de France 153
electricity 5
  system 5
ElectroCell 145
electrolysers 119, 203
electrolysis 61, 62, 152
  and storage, on site 62
  high-temperature 153
electronics 169
embrittlement 109
embryology 50
emissions 3
ENEA (National Agency for Energy, Environment and 168
ENEL 169
energy
  carrier 203
  Centre, Newcastle 142
  cleaner 40
  consumption 26
  sources,
    non-renewable types 25
    renewable 28
  units of 206
  White Paper 180
Engineering Advancement Association of Japan 116
entropy 22
  reduction 73
Environment Canada 146
enzyme 67
epoxy resin 73
Epstein, Paul 19
eruption, Lakagigar 21
Escher, Bill 132
Essen 133, 156
ethanol 34
  reforming 144
EU Transport Policy, White Paper on 151
EURO-HYPORT project 164
Europe 4, 21, 114, 150

Eastern 115
European
  Commission 109, 150-2, 155, 164, 167
  Energy and Transport Trends for 2030 114
  Joint Technology Initiative 151
  Union 34
explosion 52
Exxon 186
Eyemouth 182

F

Faraday, Michael 86
Faridabad 167
FC-Cubic 171
Fe-hydrogenase 58
Federal
  Foundation for the Brazilian Research 145
  Ministry of Economics and Technology 155
  Ministry of Economy and Labour 122
  Ministry of Education and Research 155
  Ministry of Transport 155
  University of Rio de Janeiro 145
fermentation 58
Fiat
  Panda 169
  Powertrain 169
Filbee, Sara, 138
Finland 21, 116, 176
Fischer-Tropsch 16
Fjermestad-Hagen, Elisabet 134
flame speed 84
flammability 129
Florence 168
FOI Swedish Defence Research Agency 119
Ford 85
  Motor Company 110, 182
Forschungszentrum Juelich (FZJ) 122
fossil fuels 146
  hydrogen from 54
Foundation for Research Science and Technology 142
Fourneyron, Benoit 32
FP 6 150
FP, Framework Programmes of EU, 150-152
France 5, 27, 114, 138, 152
Franklin, Benjamin 39
Fraunhofer institute for solar energy systems (FhG 122
FreedomCAR Partnership 182
Freiburg, Technical University of 82
French
  Hydrogen Association 152
  Revolution 21

fuel cell(s), 5, 24, 87, 91, 204
  alkaline 97
  applications 94
  commercialisation 146
    conference of Japan 117
    roadmap 146
  Cutting Edge Research Centre 171
  development of 86
  direct 92
    methanol 101-3
  discovery of 86
  indirect 92
  menu 90
  molten carbonate 98
  PEM 91, 94
  phosphoric acid 96
  power 172
  proton exchange membrane 94
  solid oxide 92, 99
  technologies Ltd 147
  types of 93
  UK 180
fueling station 5
Fujishima, A.K. 66
fullerene-derived nanotubes 77
Fundy, Bay of 38
fusion 28
FutureGen Industrial Alliance 185

G

gallium arsenide 66
Garman, David 137, 182
gas
  industry, Japanese 110
  natural 25
  storage 71
gasification 35, 66
gasohol 34
Gaudernack, Bjørn 173, 174
Gaz de France 153
Gazprom 179
Geels, Frank 107
Geilo 173
Geir H. Haarde 165
Gemini programme 90
General
  Atomics 60
  Electric 88
  Motors 53, 117
    Corporation 182
Generation
  II buses 184
  III phase 184
geographic information system 112
Georgetown University 184
geothermal 5
  activity 37
  energy 35
  vents 69
German 154
  Aerospace Centre (DFVLS) 122, 129
Germany 16, 31, 109, 114, 154-6, 181
GGE 204
Gibbs 88
  energy 88
Gibbs, Josiah Willard 85
Gibraltar 3-4
Ginnunga Gap 11
Girard 39
Gislason, S, 44
glacial rivers 158
glaciers 4
Glamorgan, University of 181
Global Environment Facility (GEF) 145
gluons 51
GM 170
Gore, Al 22
Gracefield Research Centre 144
graphene 77
Great Depression 21
Greece 49, 114
green algae 12
Green Initiative for Future Transport (GIFT) 166
green-house
  effect 17
  roof 13
Greenland 4, 21
  icecap 3
Gretz brothers 160
Griffith University 142
Grimsson, Olafur Ragnar 165
Grimstad 173
Grove Fuel Cell Symposium 180
Grove, William 180
Grove, William Robert 86
Grubb, William Thomas 88
GS Fuel Cell 172
Gujarat 34
Gulf of Mexico 4
Gulf Stream 4, 21
Gulf war 16
Gutowska 82
Gutowski 82

H

H2 204
H2IT 169
H2PIA 176
H2S 36, 69
Haarde, Geir H. 165
Haldor Topsoe 176
Halldór Ásgrímsson 165
Hamburg 109, 154, 160
  Society for the Introduction of Hydrogen I 160
Handy, Susan 112
Hannes Jonsson 165
Hannover Messe 156
hard policy 107
Harris, Rex 181
Hart, David 181
Harvard Medical School 19
Hauback, Björn 173
Haugesund 64
Hawaii 18, 125, 186
  University 186
heat engine 24, 204
  Carnot 92
heating value HHV 62
  and LHV 204
Heimaey 125, 126
Heimdal 43
Helion Fuel Cells 154
heliostats 30
helium 11
Helsinki
  Technical University 174
  University of Technology 176
Henan Province 25
Herning 176
Herodotus 49
high level group 151
High Temperature
  Electrolysis (HTE) 63, 153
Hindenburg airship 130
Hino 117, 170
Hjalmar Arnason 161, 162, 164
Hjalti Pall Ingolfsson 162
Hjörvarsson, Björgvin 177
Hoagland, William 135
Hobart 142
Holland 100
Home Energy Centre 182
Honda 117, 170
Honda, K. 66
Honolulu 133
HOT ELLY 63, 154
hot spot 157
House Science Committee 182
household waste 57
Hubbert, M. King 15
Huber 60
Humphry Davy 50
hurricanes 19
Husavik 36
Hussein Abdel-Aal 132
Hvalfjördur 126
Hviid, Claus 176
HyApproval 110
hybrid(s) 24, 123
  development 123
  Electric Vehicle 204
Hydrazine 91
hydrides,
  complex 80
  for Energy Storage 173
  tanks 5
hydrocarbons 13
hydroelectric
  energy 158
  power 31
hydroelectricity 146
hydrogen 3-5, 204
  atom 77
  biological methods of producing 58
  biomass, from 57
  bio resources, from 59

Index

burning of 84
chemical processes, from 54
codes 124
Carbon Containing Materials, from 134
Coordination Unit (HCU) 180
Demand, on 82
demonstration programmes 184
density of liquid 203
economy 204
    Miami Energy Conference 131
education 121
embrittlement 129
Energy
    and Fuel Cells - A vision of our future 151
    economy 5, 6
    for the Future of New Zealand 143
entering society 107
for kids 117
fossil-fuels, from 54
Highway project 147
Implementing Agreement 134
Innovation & Research Centre 176
isotopes of 51
liquefaction 204
liquid 5
metallic 53
molecules 41
nuclear energy, from 60
Park Project 168
Pathways 111
phase diagram of 53
photoelectrochemical 67
Posture Plan 183
producing 6
production methods 55
production, properties of 52
renewables, from 6
Romantics 131
safety 124, 127
solar 65
standards 124
storage compounds, exotic 82
storing 71
sulfide 56, 69
System Laboratory (HySyLab) 168
tourism 165
use 5
utilisation 84
Village project 147
waterphotolysis, from 134
wind, from 64
Hydrogenics Corporation 146
HYDROGENIUS 171
hydronium ion 89
HYDROSOL 67
HYFLEET CUTE 110
HyICE 85
HyNor 175
HySociety 117
HYSOLAR 154
Hythane 58
HYTREC (Hydrogen Technology Research Centre) 175
HyWays 114, 115

I

IAHE 132, 204
Iberia peninsula 3
ICE 204
Ice Age 13
ice caps 18
ICE hydrogen vehicles 118
Iceland 3-4, 21, 35, 37, 110, 118, 157-8, 164
    Government of 159
    University of 70, 124, 157, 159, 161, 165
Icelandic
    Alloys Ferrosilicon 125
    hydrogen project 157
    New Energy 118, 164
IEA HIA 204
Illinois University 186
Imperial College London 181
Implementation and Liaison Committee 137
Incheon 171
India 34, 39
indirect
    fuel cells 92
    policy 107
Indonesia 38
Industrial
    Research Limited 143
    Revolution 13
Industry Canada 146
INET institute 148
infrastructure 108, 124
Ingimundur Sigfússon 161
Ingolfsson, Hjalti Pall 162
Ingolfur Arnarson 160
INL (Idaho National Laboratory) 185
Institute of High Temperatures 179
Institute of People's Economy Prognostication of 179
Institute of Transportation Studies 111
insurability 131
Integrated Systems 135
    Evaluation 134
Inter-metallic compounds (IMC) 79
Intergovernmental Panel on Climate Change, IPCC 18, 22
internal combustion engines 156
International
    Association for Energy Economists 122
    Association for Hydrogen Energy 132
    Atomic Energy Agency (IAEA) 27
    Centre for Hydrogen Energy Technolog 122
    Energy Agency 26, 134
    Hydrogen Implementatio 134
    Hydrogen Energy Association 131
    Journal of Hydrogen Energy (IJHE) 133
    Partnership for the Hydrogen Economy 131, 135, 165, 183
ions 10
IPHE 204
    countries 137
    demonstration project Atlas 138
IR losses 63
IRD 65
iron 11, 60
    oxide 66
Islay, Isle of 39
isotope(s) 204
    of hydrogen 51
Israel 36
Italian
    Hydrogen and Fuel Cell Association 169
    Ministry of Environment 169
Italy 114, 167, 173

J

Japan 21, 39, 94, 116, 151, 169, 171, 181
Japan Hydrogen and Fuel Cell Demonstration (JHFC) 170
    Project (JHFC) 116
Japanese gas industry 110
Johnson Matthey Company 83
Joint Technology Initiative 152
Jon Bjorn Skulason 162
Jones, Martin 83
Jonsson, Hannes 165
Jonsson, Örn D. 165
JOULE-II project 173
Joule-Thomson
    expansion 73
    isenthalpic expansion coefficient 130
JSC "Plastpolymer" 179
Jules Verne
    Award 133
    method 61
Jülich, Projektträger 155
Jupiter 53

K

Kalina, Alexander 36
Karahnjukar 32
    dam 33
Karlsson, Erik B. 177
Katrina 15
Kawasaki 117
Keflavik 125
Kelvin temperature scale 204

KIA Motors, 172
Kloed, Christopher 173, 174
Koizumi, Junichiro 169
Konstantin Tsiolkovsky Award 133
Korea 171, 172
    Electric Power Corporation (KEPCO) 172
    Republic of 171
Krafla 69
    volcano 21-2
Krakatoa 125
Kreysa, Dr. 160
Kurchatov Institute 178, 179
Kvant RAF 177
Kyoto 151
    protocol 159
Kyushu 170
    University 171

L

Lackner, Klaus 44
LACTEC (Institute of Technology and Development) 145
Laertus, Diogenes 49
LaH2 79
Lakagigar eruption 21
Lake Charles 72
Lakehurst 130
Langer, Charles 87
LaNi$_5$H$_6$ 79
LANL (Los Alamos National Laboratory) 185
lanthanum 79
Laos 32
laptop computer batteries 95
Larderel, Francesco 35
Larminie, J. 104
Larminie, James 142
latium 37
Laughead, John, 138
Lavoisier, A. 50
Lawrence Berkeley National Laboratory (LBNL) 186
Lawrence Livermore Laboratory 53
laze 125
LBST 154
leakage 109
Leningrad 177
Leocippus 50
Lesbos 37
LH2 205
LiBH$_4$ 80
life-cycle
    analysis 54
    impacts 163
light periodic table 75
lightning 39
Limoges 153
Lincoln´s Inn 86
Lindblad, Peter 134, 174
Linde 73, 169
Lipman, Tim 112
liquefaction 73
liquid

hydrogen 5
    density of 203
    storage 156
sodium 27
storage 73
Lisbon 5
LLNL (Lawrence Livermore National Laboratory) 186
Lombardy 168
    Region of 169
London 3, 109, 182
    Olympics 182
    Stock Exchange's AIM market 181
Los Alamos National Laboratory 184, 185
Loughborough University 181
Louisiana 15
lower heating value (LHV) 62
Ludwig-Bölkow Systemtechnik 114
Lucchese, Paul, 138
Lund, Peter 174
Luxembourg 109
Luzzi, Andreas 134
Lyon 133

M

M. King Hubbert 15
Maack, Maria 118, 162, 165
Madrid 109
Maeland, Arnulf J. 173
Magdalene College 85
Magma 158
magnetic
    field, Earth's 10
    refrigeration 74
Majzoub, Eric 81
Malcolm Wicks 180
Malyshenko, Stanislav 177
MAN 85
Mantova 168
Mao, Zong-Qiang 148
Marchetti, Cesare 131
Marco Polo 25
Marghera 168
Maria Maack 118, 162, 165
Martinez, Anibal 132
Massey University 143
Massif Central 37
Mauna Loa Mountain 18
Mawson base, Antarctica 142
Max-Planck Institute 80
McCain, John, 183
MCFC 205
Meadi 30
Melbourne 142
Messina 168
metal
    crystal, lattice of 77
    hydride storage 75, 205
    organic-framework compounds 83
    ruthenium 4
metallic hydrogen 53

meteorites 12
methane 14, 127
methanol 56, 95
METI 169
Mezen Bay 38
Mg$_2$FeH$_6$ 80
MgH$_2$ 78
Miami 133
Michael Faraday 86
Michel 60
Michigan, University of 83
Milan 168
Miletus 49
Millennium Cell 81
Mills, Mike 138
Milwaukee 88
Ministry of Commerce, Industry and Energy (MOCIE) 171
Ministry of Economic Development 142
Ministry of Economy, Trade and Industry (METI 169
Ministry of Economy Trade and Industry, Japan 116
Ministry of Mines and Energy 144
Ministry of Non-Conventional Energy Sources 166
Ministry of Science and Technology 144, 149
Ministry of Science and Technology (MOST) 171
Mitchell 53
Mitsubishi 117, 170
Mogensen 64
Mojave Desert 30
Mok, Phillip 162
molten carbonate 91-2
    fuel cell (MCFC) 98, 172
Monash University 142
Mond, Ludwig 87
Montecatini 167
Montreal 133
Moon 10
Morocco 4
Moscow 133, 177
    motor show 178
Moses 14
motor show, Moscow 178
Mouchout, Auguste 30
MTU CFC Solutions 98
Murdoch University 142
Mysterious Island, the 61, 160

N

Nafion 90, 93-4, 102
Nancy 153
Nanostructured carbon 77
Nanotube 77
NASA (National Aeronautics and Space Administration) 74, 182
Nathan Parker 113
National
    Diet of Japan 169
    Fuel Cell Bus Technology

Index

Development Prog 184
    Hydrogen
        Commission 174
        Energy Board (NHEB) 166
        Energy Road Map 166
        Hydrogen Institute of Australia 141
        Hydrogen Roadmap 183
        Innovation Programme on Hydrogen and Fuel 155
        Research Council (NRC) 146, 183
        Solar Energy Centre 142
Nationale Organisation für Wasserstoff und Brennst 155
NATO 125
Natrium 82
natural gas 3, 25
    power system 43
    reforming 144
Natural Resources Canada 146
Neef, Hanns-Joachim 138
Nellis 53
Nesjavellir plant 35
Netherlands 31, 109, 114
NETL (National Energy Technology Laboratory) 186
neutrons 27
Nevada-Reno University 186
New Business Venture Fund 160
New Delhi 167
New Energy and Industrial Technology Development 170
New England Journal of Medicine 20
New Haven 85
New Jersey 130
New South Wales, University of 142
New York 148
    Mercantile Exchange 16
New Zealand 35, 141-2, 144
Newcastle, UK 25
Newcastle, Australia 142
Nicholson 87
Nicholson, William 61
nickel
    hydroxide 79
    metal hydride battery 80
    cadmium 79
Niedrach, Leonard 90
NiFe hydrogenase 58
Nikolaus Otto 15, 23
Nissan 117
nitrogenase 58
nitrous oxides 17
NMH battery 79
Nobel
    Peace Prize 22
    Prize 29, 60
non-renewable 25
Nordic 172
    countries 119
    Energy Research 176
        Programme 174

Hydrogen Energy Foresight 119
Noreus, Dag 173
Nornikel 179
Norsk Hydro 64, 161, 172, 174-5
North Africa 4, 5
North Sea 3, 43, 174, 176
North Vancouver 147
Northern
    Europe. 5
    Lights 9
Northrhine-Westphalia 154
Northumbria, University of 181
Norway 4, 39, 43, 114, 118, 173-4
Norwegian Hydrogen Strategy 174
notable surface of action 93
Nottingham, University of 181
NREL (National Renewable Energy Laboratory) 186
nuclear
    energy 27, 146, 153
    fission 27
    fusion 28
    hydrogen from 60
    Ministry (Minatom) of Russian Federation 179
    Hydrogen Programme 172
    plants 42

O

Obadiah Bartholomy 113
ocean
    currents 4
    Power Technologies 39
    Thermal Energy Conversion (OTEC) 38
octahedron 78
Oddsson, David 165
Ogden, Joan 111-4
Ohta, Tokio 132
oil 14, 157
    platforms 3
    production 16
    window 14
Olafsson, Sveinn 165
Olafur Ragnar Grimsson 165
Olympic Games 147
    Beijing 2008, 150
    London 2012 182
Omar Yaghi 83
Omni Shoreham Hotel 135
OPEC 16, 132
orbiter power plant 90
Orissa 34
Ormat Inc 36
Örn D. Jonsson 165
ORNL (Oak Ridge National Laboratory) 186
ortho- hydrogen 73, 205
Oslo 175-6
Össur Skarphédinsson 165
Ostwald, W. 88
Otto Cycle 23
Otto, Nikolaus 15, 23

Ottobrunn 114
Owen, Anthony 122
Oxford, , University of 181
Oxford, University of 83
oxidation, partial 56
Oxygen 12
ozone 17

P

P-Primacy of Policies 120
Pacific Northwest National Laboratory 82
Pacific Ocean 18
PAFC 205
Pakala 41
Palermo 168
Pall Kr. Palsson 160, 162
Panasonic 169
Panick, Dr. 161
Pannonian Basin 37
para-hydrogen 73, 205
parabolic dish system 30
Paralympics Winter Games 147
Paris 50, 55, 127, 133, 160
Parker, Nathan 113
partial oxidation 56, 205
Pasadena 133
Pascal 205
Patchkovskii 82
PdH0.6 78
peat 157
PEFC 205
PEM 91
    fuel cell 91, 94
PEMFC 205
periodic table, light 75
Perth 109, 141
Petersen, Jan Hovald 176
Petrobras 145
petrol 13-4, 42
Peugeot Citroën 153
phase diagram 52
Philippines 38
Philosophical Magazine 87
PHOEBUS 154
phosphoric acid fuel cell (PAFC) 91, 96, 184
photobiological
    production, 134
    water splitting 205
photodecomposition 205
photoelectric effect 29
photoelectrochemical hydrogen production 67
photoelectrochemical system (PEC) 66
Photoelectrolytic Production 134
Photons 17
photosynthesis 12, 59
photovoltaic(s), 28
    production 42
physiosorption 76
Piedmont 168
pipelines 109

Pittsburgh 186
　University 186
plasma
　phase 52
　technology 28
Plasmachemical technologies 178
plaster of Paris 87
platinum 103, 123
Poetic Eddas 11
Poland 116
polar ice 3
Pole star 9
policy targets 107
polyaminoborane 82
Poncelet, Victor 123
Porto 109
Portugal 39, 109
power plant(s), 4
　decarbonisation of 42
Power Tower System 30
Precombustion Decarbonisation 134
Preis, H. 90
Prince Edward Island 148
Princeton 114
　University 41
Prodi, Romano 150
production methods, hydrogen 55
Programme on Investment into the Future (ZIP) 155
Projektträger Jülich 155
protium 51, 205
proton(s) 10, 50, 93, 205
　discovering the 49
　exchange membrane 92, 93
　　fuel cell 93-4
protonics 169
Prout, William 50
public acceptance 163
purple bacteria 60
Putin, President 179
pyrolysis 34, 57

Q

Quad 26
Quadrillion Btu 26
quantum mechanical tunnelling 28
quarks 51
Queensland, University of 142

R

Raadhuspladsen 131
Radioactive waste 27
Raldow, W. 134
Raman spectroscopy 81
Rance estuary 38
RAO 179
RAUDI 57
Raufoss Fuel Systems 175
Reading University 181
reforming 66
renewable(s) 25, 57, 206
　energy 205
　　sources 28

hydrogen from 57-67
Research Centre for Hydrogen Industrial Use 171
Research Centre Jülich 154
Rex Harris 181
Reykjavik 9, 55
　Energy Corporation 36
Rhine Graben 37
Richmann, Georg Wilhelm 39
Richmann, Wilhelm 39
Riis, Tryggve 134, 174
rings of fire 37
Rio de Janeiro 145
　Federal University of 145
Risö
　Institute 65
　National Laboratory 119, 173, 176
Riyadh 154
Roadmap for Manufacturing R&D on the Hydrogen Econ 184
Roskilde 65
Rossmeissl, N. 134
Rudolph A. Erren Award 133
Ruhr 72
Russia 15, 90, 177-9
Russian
　Academy of Sciences 178
　Aviation & Space Agency 179
　Federation 177
ruthenium metal 4, 68
Rutherford Appleton Laboratory 83, 181
Rutherford, Ernest 50, 51

S

Sachs, Jeffrey 44
Sachs, Rene 152
Sacramento 111
　County 112
　Municipal Utility District 113
Sætre, Tor 174
SAFETEA-LU 184
safety
　code officials, and 121
　hydrogen, of 135
Sahara 3
Salerno 168
Samsung 172
Sandia Laboratory 81, 185-6
Sandrock, Gary 83
Sao Paulo 144
SAPHYS Project 173
Saudi Arabia 15, 154
Savannah River National Laboratory 186
Scandinavia 172
Schleisner, Lotte 174
Schnieder, Harald 162
Schoenung, Susan 135
Schönbein, Christian F. 87
Schrieber, G. 134
Schuman, Frank 30
Schwarzenegger, Governor 114

Schwickardi, M. 80
Scotland 39
Scottish Hydrogen and Fuel Cell Association 182
Sebkha City 5
sedimentary deposits 18
Seifritz, Walter 132
Senju 117
Seoul 171
sequestration 108
　programme 4
Severn Estuary 38
sewage sludge 57
Shanghai Sun Li High Technology Company 149
Sheffield Scientific School 85
Shelish, B.I. 177
Shell 15, 186
　Hydrogen 110, 161
Sicily 4
Sigfússon, Ingimundur 160, 161
Sigfusson, Thorsteinn I 164-5
Sigurgeirsson, Thorbjorn 124
silicon carbide 96
silver
　chloride 66
　iodide 89
Simon, F. 73
Simonnet, Antoine 113
Sir William Grove Award 133
Skarphédinsson, Össur 165
Skogsholm, Torid 174
Skulason, Jon Bjorn 162, 164
Social Sciences and Humanities Research Council 146
society 107
society, hydrogen entering 107
socioeconomic 107, 108
sodium
　borohydrides 165
　metaborate 82
Soerensen, Bent 176
SOFC 206
soft policy 107
soil management, agricultural 43
Sokolow, Robert 41, 43
solar
　energy 4, 5
　hydrogen 4, 65
　　conversion paths 66
　power 30
　　concentrating 30
　　systems, concentrated 30
　thermal 65
　thermochemical 66
　wind 10
solid
　and liquid state hydrogen storage materials 135
　　storage 134
solid
　oxide fuel cells 91-2, 99, 146
　state storage 75
Sophia-Antipolis 153
sorption properties 76

Index

South Africa 16
South Wales 182
Soviet Union, former 16
space rockets 52
Spain 31, 109, 116
Special Integrative Fund for Research (FISR) 168
Spencer Abraham 135
Sperling, Dan 112, 114
spin 73
St. Malo 38
Standards, Codes and 130
Stanford University 186
START 109
State University of Campinas 145
static electricity 130
Statkraft 176
Stavanger 175
Steacie Institute for Molecular Sciences 82
Steady State City Hydrogen System Model (SSCHSM) 112
steam reforming 55, 206
Steering Committee 137
Stern, Sir Nicholas 21
Stirling engine 65
Stockholm 109
Stockholm University 173
storage,
    gas 71
    liquid 73
    metal hydride 75
    on site electrolysis and 62
    solid state 75
    hydrogen 71
Strategic
    Plan 183
    Research Agenda and Deployment Strategy 151
Strathclyde University 181
strings 11
Strongsil 126
Stuart, A.K. 134
Stuart Energy 146
students 121
Stuttgart 109, 133, 154, 160
substratum 50
sugar cane 144
sulfonated polystyrene 89
sulfur 100
    dioxide 34
    iodide 60, 65
Sun, The 9-10
Sund, Bjorn 162, 173
SUPERGEN programme 180
surface temperature, Earth's 20
Surrey 147, 148
Surtsey 124
Sustainable Marine and Road Transport, H2 in Iceland 165
Sustainable Transport Energy Perth 142
Suzuki 170
Sveinn Olafsson 165
Sverrisdottir, Valgerdur 162

Swansea 86
Sweden 4, 37, 109, 119, 176
synchrotron 83
Syria 131

T

Tampere, Technical University of 176
tanks, hydrides 5
Tarfaya 5
Tasmania, University of 142
teachers 121
Technical University
    Freiburg 82
    Tampere 176
Technological Institute 165
Technology Partnership Canada 146
tectonic
    boundaries 157
    plates 37
Tees Valley 72, 109
Teflon 90
tellurium 27
temperature,
    Earth's surface 20
    production of hydrogen, high and low 134
testing forum 157
tetrahedron 78
Thales of Miletus 49
Thames 3
thermal units (British) 26
thermochemical 57
    cycles 66, 152
    water splitting 61
thermodynamics 22
thermoelectricity 39
thermolysis 66
thermophilic anaerobic bacteria 58
Thierry Alleau 152
Think Nordic AS 175
Thomson, J.J. 50
Thorbjorn Sigurgeirsson 124
Thorolfur, Arnason, 164
Thorstein Vaaland 173
Thorsteinn I Sigfusson 164-5
Three Gorges 33
three wheelers 166
tidal power 38
TiFe 75
Tim Lipman 112
titanium dioxide 66
Titusville 14
Tokio Ohta 132
Tokyo 133
    Bay 116-7
Toronto 133, 147
    Mississauga, University of 147
TOTAL energy company 113
Totara Valley 144
Towards a National Hydrogen and Fuel Cell Strategy 146
town gas 55
Toyota 117
    Prius 24

Transactions 86
Transitional Hydrogen Economy Replacement Model 112
Transport Canada 146
transportation infrastructures 108
Treptow 153
tritium 51, 206
trough System 30
Tse 82
Tsinghua University 148
Tsiolkovsky, K.E. 177
TU 155 aircraft 177
tungsten trioxide 66
Tunold, Reidar 173
Turin 168
Turkey 132
Turkish Ministry of Energy and Natural Resources 122
Tuscany 37, 168
Twente, University of 107

U

U.S. Geothermal Energy Association 38
UK 28, 116, 138
    Fuel Cell Development and Deployment Roadmap 180
    Hydrogen Association 182
    Research Council 180
Ulm 122
UMIST 181
Unitech 145
United Energy Systems 179
United Kingdom 109, 180
United Nations 6, 122
    Development Programme 145
United States 5, 15, 27, 31, 39, 94, 151, 181, 186
    Council for Automotive Research 182
United Technologies Corporation 90
units of energy 206
University of
    California at Davis 110-11
    Iceland 70, 124, 157, 159, 161, 165
    Michigan 83
    New South Wales 142
    Oxford 83
    Queensland 142
    Tasmania 142
    Toronto at Mississauga 147
    Twente 107
Uppsala University 174, 177
upstream 43
Uranium 27
Urey, H.C. 50
US National Hydrogen Association 186
UT-3 60
Utah 186
    University 186
Utsira (Utility Systems in Remote Area) 64, 174

V

Vaaland, Thorstein 173
Valgerdur Sverrisdottir 162
Van Allen Radiation Belts 10, 206
Van der Waals 76
Vancouver 147, 148
    Fuel Cell Vehicle Programme 148
Vartsilla 176
VAZ automobile 178
Vega, Louis 38
Veneto 168
Venezuela 132
Vestmannaeyjar 124
Veziroglu, Nejat 122, 131-2
Victoria 147, 148
Vienna 133
Viking time heritage 64
Vilhjálmur Vilhjálmsson 165
Vinnova 177
Vision of America's Transition to a hydrogen Economy 183
volatile components 57
volcanic eruptions 19
volcano, Krafla 21
Volga car 177
Volta, Alexandro 86
Volvo 85
Vorst, Bill van 132
VTT
    institute 119
    Technical Research Centre 176

W

Walker, Robert 182
Washington DC 135, 182, 187
waste heat 39
water 12
    electrolysis 144
    electrolysis of 61
    gas shift reaction 55
    management 103
    splitting, photobiological 205
waterwheels 23
Watt, James 23
wave
    and tidal power 38
    power 38
WaveGen 39
weather events, costs of extreme 20
wedges 42
Weil, Kurt 132
Weinert, Jonathan 111
Weir 53
Well-to-Wheel Efficiency 206
Wellington 144
Western Anatolia 37
Western Australia 142
Westphal, C. 88
wheat 34
Whistler 147
White Paper on EU Transport Policy 151
Wicks, Malcolm 180
Widvey, Thorild 174
Wigner, Eugene 53
wind 5, 64
    energy 4, 31
    generators 3
    hydrogen from 64
    solar hybrid systems 5
windmills 23
Winter, Carl-Jochen 5, 154, 156-7
Winter, Eva, 157
Wisconsin 88
Wolf 82
Wolfgang Dönitz 154
World
    Bank 36
    Demonstration Atlas 138
    Hydrogen Energy Conferences (WHECs) 133, 157
    Trade Centre 148
    War II 88
    Wars 21

X

Xerxes 14

Y

yachts 5
Yaghi, Omar 83
Yakushima 170
Yale University 85
Yartys, Volodymyr 173
Yexian County 25
Yokohama 117, 133

Z

Zaire 62
Zero Regio 168
    Project 168
zinc
    oxide 67
    sulfide 56
zirconium 27
Zong-Qiang Mao 148
Zorbas, Stephen 141
Zurich 133
Zweig, Bob 132